Transparency 2.0

Susan Drucker
General Editor

Vol. 3

The Communication Law series is part of the Peter Lang
Media and Communication list.
Every volume is peer reviewed and meets
the highest quality standards for content and production.

PETER LANG
New York • Washington, D.C./Baltimore • Bern
Frankfurt • Berlin • Brussels • Vienna • Oxford

Transparency 2.0

Digital Data and Privacy
in a Wired World

Edited by Charles N. Davis
and David Cuillier

PETER LANG
New York • Washington, D.C./Baltimore • Bern
Frankfurt • Berlin • Brussels • Vienna • Oxford

Library of Congress Cataloging-in-Publication Data

Transparency 2.0: digital data and privacy in a wired world /
edited by Charles N. Davis, David Cuillier.
pages cm. — (Communication law; vol. 3)
Includes bibliographical references and index.
1. Data protection—Law and legislation—United States.
2. Freedom of information—United States. 3. Digital media—Law and legislation—
United States. 4. Social media—Law and legislation—United States.
I. Davis, Charles N., editor of compilation. II. Cuillier, David, editor of compilation.
KF1263.C65T734   342.7308'58—dc23   2014005847
ISBN 978-1-4331-1744-2 (hardcover)
ISBN 978-1-4331-1743-5 (paperback)
ISBN 978-1-4539-1333-8 (e-book)
ISSN 2153-1390

Bibliographic information published by **Die Deutsche Nationalbibliothek**.
Die Deutsche Nationalbibliothek lists this publication in the "Deutsche
Nationalbibliografie"; detailed bibliographic data is available
on the Internet at http://dnb.d-nb.de/.

© 2014 Peter Lang Publishing, Inc., New York
29 Broadway, 18th floor, New York, NY 10006
www.peterlang.com

All rights reserved.
Reprint or reproduction, even partially, in all forms such as microfilm,
xerography, microfiche, microcard, and offset strictly prohibited.

Table of Contents

Introduction vii

**Part One: Battlefield Court:
     Balancing Access and Privacy** 1

Chapter One: The 'Practical Obscurity' Doctrine:
     When Is a Public Record Too Public? 3
     Sigman L. Splichal

Chapter Two: Tipping the Scales: How the U.S. Supreme Court Eviscerated
     Freedom of Information in Favor of Privacy 16
     Martin E. Halstuk, Benjamin W. Cramer, & Michael D. Todd

Chapter Three: Public Access and Informational Privacy
     in Electronic Government Databases 36
     Joey Senat

Chapter Four: Conflict in a Digital World:
     The European Context 51
     Cheryl Ann Bishop

Part Two: Online Dilemmas: Email, Social Media, and Those Pesky Jammie Surfers ... 65

Chapter Five: Electronic Court Record Access: Present Landscape, Neutral Principles, and the Looming Interloper of Contextual Privacy ... 67
Richard J. Peltz-Steele

Chapter Six: Social Media and Reporting on Judicial Proceedings: A Digital Era Conflict ... 84
Derigan Silver

Chapter Seven: Access to Email and the Right of Privacy in the Workplace ... 97
Kyu Ho Youm

Part Three: Looking Ahead: Leaker Chaos, Resolution, and the Millennial Shift ... 115

Chapter Eight: All the News That's Fit to Leak ... 117
Jonathan Peters

Chapter Nine: Finding Resolution: Systems for Resolving Disputes and Reconciling Access with Privacy ... 133
Daxton R. "Chip" Stewart

Chapter Ten: Here's Looking at Me: The Abandonment of Privacy and Solitude as Millennials Move to Life Online ... 145
Paul Gates

List of Editors and Contributors ... 161
Index ... 167

Introduction

Few modern legal controversies engender as much passion, or as much confusion, as the unavoidable conflict between the protections for personal privacy and the right of access to information kept by government agencies about those very same people. Add technology and nearly unrestrained digital information flow to the mixture, and the world becomes a whirlpool of intrusion, empowerment, confusion, and fear. The headlines blare Wikileaks, Edward Snowden, National Security Administration wiretapping, big data, and identity theft, causing citizens to recoil, privacy defenders to mobilize, access advocates to fret, and judges to sort out an appropriate balance. The nexus of access and privacy remains largely unsettled, a paradoxical legal swamp devoid of concrete doctrine, relying instead upon a patchwork quilt of ad hoc judicial balancing, fitful attempts at rulemaking, and statutory pronouncements that all serve to create a dizzying array of unpredictable outcomes. Add the modern propensity for commentators to lump all privacy issues together, despite their very distinct legal genealogies, and the unsurprising result is a body of law shaped in large part by fact-bound judicial pronouncements that offer little in the way of guidance for policymakers and legislators seeking coherence with access principles.

Freedom of information law—the body of law emanating from right-to-know statutes in some regimes, and from constitutional rights of access in others—emanates from the democratic principles of self-governance and accountability. To allow the citizenry to monitor its government, it follows logically that a presumption of openness must undergird the records and proceedings of its agencies.

In the everyday business of governing, agencies must as a matter of course collect a great deal of information about citizens. From the mundane licensure and registration of businesses, real estate transactions and the like to the much more sensitive collection of law enforcement information, health care data, and other highly intimate, personal information, millions upon millions of records fall within the ambit of government.

The emergence of digital record keeping introduced an entirely new range of issues to an already chaotic area of access law. Seemingly overnight, records once stored in relative obscurity in government file folders became available with the press of a few buttons. Laws created to provide access to paper records were ill equipped to negotiate the questions wrought by digital access. Records seen by few suddenly were available to everyone, in real time, raising new issues and generating conflicts that strike at the core of the presumption of openness.

The scope of government record keeping made possible by digital technologies, and the ease with which records can be produced, duplicated, and disseminated, offer both promise and peril. The potential for digital records to transform participatory self-governance knows no bounds, and allows the citizenry to organize and engage in ways unthinkable a generation ago. Instantaneous communication and web-based dissemination of government information provides tools that empower citizens and promises to level the playing field between the governed and their governors, ushering in a new era of government accountability.

That promise, unfortunately, carries with it the prospect of very real threats to personal privacy, as digital technologies concurrently make it easier than ever for prying eyes to misuse information even as a greater proportion of the information kept by the state on the lives of individual citizens finds its way online.

Reconciling that great legal conflict—the inevitable clash between openness and privacy—forms the basis of this book, which treats several of the most pressing issues of the day as a means for exploring the current state of informational privacy, and to where it might lead in the future.

The conflicts highlighted in this work represent but a few of the vexing issues arising from digital privacy. Indeed, discussions of informational privacy almost inevitably run aground as people tend to conflate one privacy arena from another. Our students constantly struggle with separating informational privacy from personal privacy from reproductive privacy, despite the fact that each brings with it an entirely different progeny of law, each with its own linchpins and definitional certainties.

It is deceptively easy to describe the ultimate conflict between privacy and access as a negotiation between the doctrinal principles of each body of law: the presumption of openness that forms the basis of access law versus the expectation of privacy giving rise to informational privacy. While such a formulation historically has served to inform the judicial balancing inherent in the case law, it provides little guidance for emerging conflicts like Wikileaks and the Snowden affair, which represent a new phenomenon in which technology gives rise to massive disclosures of government secrets. Informational privacy doctrine has little to offer in the way of reconciling these disclosures.

Little about either incident finds comfort in traditional privacy law. Snowden and Julian Assange each defended their disclosure in terms of defending privacy from unwarranted government intrusion. Each pits governmental assertions of the value of secrecy against broader societal interests in public knowledge of governmental actions, an age-old formulation of the benefits of access, yet neither considered the privacy rights of individuals harmed by the disclosures. In many ways, these timely examples amply demonstrate the shifting nature of the interests at stake in digital dissemination. As governments worldwide embrace decentralized records systems and classify more and more documents, the unsurprising result is the emergence of leakers acting in self-defined public interest terms. The future doubtless will produce new revelations, as system-wide classification regimes collapse of their own weight.

The legal protections for privacy emerged in response to specific phenomena; in the case of informational privacy, it could be argued that the rise of the surveillance state and its media counterparts drove the legal response as much as anything else. Privacy law has always been a response to emergent forms of surveillance and information gathering, and in that context, none of the controversies discussed in this book should surprise us.

The pace which new digital tools have appeared, with each new generation doubling and redoubling the speed and scope of communication, represents the greatest challenge for the law. At the dawn of the twentieth century, the camera represented the state of intrusive technology, giving rise to the first scholarly discussion of legal privacy rights. How quaint!

The seminal work of Samuel Warren and Louis Brandeis, "The Right to Privacy," in an 1890 *Harvard Law Review* article, decried the use of the newfangled camera and other intrusions into what today would appear to be wholly public places as the foundation of individual freedom in the modern age. Given the increasing capacity of government, the press, and other agencies and institutions to invade previously inaccessible aspects of personal activity, they argued that the law must evolve in response to technological change.

In fact, the central thrust of Warren and Brandeis' article was an attempt to distinguish privacy from property rights, and to delineate some legally prohibitive harm in the act of intruding upon that elusive right. As one court remarked, after noting the work of Warren and Brandeis: "Basically, recognition of the right to privacy means that the law will take cognizance of an injury, even though no right of property or contract may be involved and even though the damages resulting are exclusively those of mental anguish."

Therein lies the great paradox of privacy law. It sets forth a right to be protected yet the right is ill defined, and its legal parameters evade firm definition. Privacy law's grounding in the amorphous "penumbras" of constitutional protection championed by Justice Brennan in *Roe v. Wade* perfectly illustrate why this area of the law can prove elusive for legal students grounded in statutory precision and three-part tests.

If we can presume that government transparency is a legal right of equal worth to privacy, or at the least is an important component of democratic self-governance that cannot simply be swept aside in the service of privacy interests, the complexity of the task before us is made clear.

As the following chapters demonstrate, digital technology has again and again altered the privacy landscape by raising novel questions that render the traditional judicial and legislative balancing of interests unable to adequately serve the interests at stake in these disputes. Several structural weaknesses exist in each of these conflicts.

First, it can be argued that digital technologies and the changes wrought by them work to elevate privacy interests above the values of transparency. This is examined throughout the book, starting in Chapter One by Sigman L. Splichal, and in Chapter Two by Martin E. Halstuk, Benjamin W. Cramer, and Michael D. Todd. The chapters explore "practical obscurity" and how, the authors argue, U.S. Supreme Court rulings have eviscerated access statutes in large part because of concerns over privacy and digital information proliferation. Joey Senat in Chapter Three and Richard J. Peltz-Steele in Chapter Five build on the argument by looking at how the courts have approached government data cautiously, nervous of the ease with which anyone can peruse court records in databases or online, easily snooping on neighbors, co-workers, and family members. Even covering public proceedings, such as trials, through social media, can cause consternation and conflicting rules when balancing privacy with access, as Derigan Silver explains in Chapter Six, not to mention how to traverse email and other online communications, as Kyu Ho Youm explores in Chapter Seven. Whether the issue is access to the emails of government officials, access to government databases, instant messaging, or other emerging forms of digital collection or dissemination of information, privacy interests enjoy an exalted status. This is not to suggest any sinister craving for secrecy. Rather, it is a natural reaction to the digital unknown, a predictable and oft-repeated response fueled by political realities. Simply put, "protecting" privacy by restricting new digital avenues of access to governmental information offers little political risk. Asked whether they desire greater privacy, few citizens respond by questioning the need for it, and fewer still demand that arguments in favor of access receive an equal place at the policy table.

Second, arguments for digital transparency often are grounded in abstract notions of "the right to know," self-governance, and accountability. All are important values, but not one is sufficiently evocative to compete with privacy narratives, which often are based on worst-case scenarios and heuristic shortcuts designed to overcome those arguments. Good government, it seems, is a tough sell in the face of powerful, if specious, privacy arguments. This is not to say that powerful and real privacy arguments do not exist; quite the contrary. In some cases even the most powerful arguments for access must yield to stronger privacy interests. This book demonstrates, however, that in many digital transparency disputes, the disciplined analysis of competing interests required to better reconcile these conflicts falls victim to political calculus designed to cater to temporal responses to new technologies. This makes resolution more difficult, as Daxton "Chip" Stewart describes in Chapter Nine regarding processes used to reach agreement and find common ground.

Third, it is important to acknowledge that informational privacy and attitudes are dependent on historical and cultural conditions that shape a society. In Chapter Four, Cheryl Ann Bishop explains the evolving concept in Europe of the "Right to be Forgotten." Many Europeans take their personal privacy very seriously, particularly protections from government intrusion given the history of the German Stasi and Russian KGB. At the same time, Europeans insist on holding their governments accountable and have some of the strongest freedom of information laws in the world, escalating the conflict between privacy and access. Global perspective is important in the digital age because information is no longer confined to file cabinets—it can spread through the world within seconds. What one country considers private, another simply the front page of a supermarket tabloid or blog. As we've seen from Wikileaks, discussed in Chapter Eight by Jonathan Peters, the ease with which digital data can travel the globe unchecked exponentially magnifies the potential to foster greater government accountability, but at the same time amplifies the threats to the privacy and safety of individuals.

Fourth, and perhaps most importantly, this book ably demonstrates throughout that informational privacy suffers from an incoherent, disorganized legal footing. The lack of any single basis for informational privacy gives rise to a bewildering mix of statutes, common law pronouncements, and state constitutional provisions that combine to create a volatility underlying the law of digital privacy. The effect of this hodgepodge of legal analysis makes it difficult to reconcile access policy in the face of new technologies. Rather than being able to turn to concrete legal principles, legal analysts and policy makers face the unenviable task of confronting seemingly immobile constants. On the one hand, access law begins with the presumption of openness: freedom of information law begins with the conceptual starting point that unless the law states otherwise, information should be public. Such a presumption should weigh heavily upon considerations of informational technologies, but as the foregoing chapters demonstrate, that presumption easily gets swept aside in the face of privacy arguments. The presumption of openness, it seems, is about as strong as any legal aphorism, yielding to a wide range of presumed harms should digital technology actually make information more available, in readily available formats, to users who in turn use that information to create new products and services.

The editors of this volume, longtime freedom of information advocates, offer no apologies for their overt bias towards freedom of information. We see the technologies discussed in this book not as sinister new forms of intrusion, but as exciting new ways to empower citizens to better understand their government and its impact on the citizenry. With the exception of Paul Gates' penetrating critique of social networking in Chapter Ten, the book stands as an affirmation of digital transparency and as a celebration of digital democracy. We welcomed Gates' analysis as a strong counterpoint to our overall philosophy that the more the transparency, the greater the good, for we readily acknowledge that ours is far from the only take on these new technological tools.

This book is a sum of its parts, an examination of many of the most pressing issues surrounding digital access to government information but by no means the last word on the subject. It is our hope that the discussion is only beginning, and that this work is but the opening salvo in a growing body of work bringing scholarly attention to these issues and many others at the nexus of access and privacy in a digital age.

PART ONE

Battlefield Court: Balancing Access and Privacy

CHAPTER ONE

The 'Practical Obscurity' Doctrine: When Is a Public Record Too Public?

SIGMAN L. SPLICHAL

American society has long recognized the value of a second chance. Many immigrants came to this nation to escape deprivations or persecutions; to start anew. Others fled to the frontiers of the new nation to distance themselves from mistakes and misfortunes.

For much of the first two centuries of American history, the natural barriers of time and distance served well those who sought new lives. These practical barriers, coupled with the inherently cumbersome nature of governments' decentralized paper record-keeping practices, ensured at least a measure of obscurity for those eluding their "official" pasts.

But new technologies began to break down these barriers, both speeding the flow of information and making it easier to manage. At the center of the information revolution was the computer. As Justice Rehnquist once noted, in words that express the concerns of many as the twenty-first century loomed: "There is no frontier...(to go) to get a new start. All of our important acts, our setbacks, the accusations made against us go into data banks and are instantly retrievable by the computer."[1]

In the late 1950s and early 1960s, the computer was heralded for its seemingly limitless capabilities, which commentators saw as a means of enhancing the quality of life and freeing people for higher pursuits. One commentator put the computer on a par with the atomic bomb as the most significant twentieth-century innovations.[2] Among the benefits of computer technology was the ability to sort and analyze vast amounts of information accumulated by various agencies within government.

But as technology became more sophisticated, and as government expanded its reach into personal lives through a myriad of social welfare and other programs, the tenor of the public discussion changed. No longer was the computer seen for its social advantages. The specter of "Big Brother" soon overshadowed the computer's advantages as fears were raised of the abuse of private information. These concerns were the focus of a series of congressional hearings spawned by a proposal by the U.S. Bureau of the Budget (now the Office of Management and Budget) for creation of a national data center—a centralized computer system in which to pool information gathered by various federal agencies. Those hearings culminated with passage of the Privacy Act of 1974, legislation designed to ensure government information practices did not infringe on personal privacy.[3] As U.S. Rep. Frank Horton of New York observed as a member of a House subcommittee involved in the national data center controversy in 1966: "Information is scattered in little bits and pieces across the geography and years of our life. Retrieval is impractical and often impossible. A central data bank removes completely this safeguard."[4]

Concerns about technology and privacy, expressed by commentators and legislators, have not been lost on the U.S. Supreme Court. Such reservations were expressed by members of the Court for a number of years, leading Justice William Brennan to caution that he was not "prepared to say that future developments will not demonstrate the necessity of some curbs on such technology."[5]

In 1989, in a seminal case in public access law, the Supreme Court focused on computer privacy concerns in an opinion that threatens to reduce access to government information. In *U.S. Department of Justice v. Reporters Committee for Freedom of the Press*, the court articulated the "practical obscurity" doctrine. It suggested that personal information—even when taken from public record sources—regained its privacy interest when compiled in government computer databases.[6]

Computers have replaced the file cabinet as the storage place of choice among government agencies and bureaucracies. Access to this information, and the laws and customs affecting such access, are major issues confronting the public and news media in the waning years of the century. Personal privacy is at the center of the access debate.

This chapter explores the U.S. Supreme Court's attempt to deal with one festering computer/access issue—the rights of individuals to control information about themselves in the computer age. The chapter looks at several early cases in which the Court grappled with the computer privacy issue, expressing concerns about technology and its potential to violate privacy. It then analyzes the Reporters Committee opinion as it deals with informational privacy. Finally, the chapter concludes that while the Court's concern about the dangers that databases pose to privacy is timely and well-intentioned, it nonetheless threatens public and media access to information by shifting the focus of analysis in access cases away from the public nature and content of the information to its computerized form.

EARLY PRIVACY TECHNOLOGY CASES

The evolution of judicial perspectives toward privacy, access to information, and computers, has spanned decades in conjunction with the development of technology. Given the U.S. Constitution contains no direct protections of "privacy," this concept has developed case by case, picking up steam at about the same time as the introduction of the Altair 8800 and other beta home microcomputers.

Paul v. Davis—No Informational Privacy

The initial reluctance of the Supreme Court to fully embrace the concept of informational privacy is illustrated in a pair of 1976 cases. In *Paul v. Davis*, a man against whom shoplifting charges were filed but later dismissed sought relief after police in Louisville, Kentucky, included his name in a listing of "active shoplifters" distributed to 800 local merchants.[7] On a 5-3 vote, the Court refused to expand constitutional privacy to cover the dissemination by police of such personal and defamatory information. Justice Rehnquist, who later became chief justice, balked at the notion of a constitutional basis for informational privacy. He distinguished privacy interests in the control of personal information from privacy cases that dealt with substantive restrictions on activities such as contraception and procreation. "None of our substantive privacy decisions hold this or anything like this, and we decline to enlarge them in this manner," Justice Rehnquist wrote.[8]

United States v. Miller—Beginnings of Discomfort

In *United States v. Miller*, decided soon after *Paul*, the Court again declined to recognize a right of information privacy in a case that dealt with a man's control of personal information he provided to a bank.[9] The Fourth Amendment case, decided on a 5-4 vote, addressed privacy interests in information given to third parties, such as businesses, not to the government. Because the government was not a party to the information-gathering and dissemination in the case, the activities of the bank did not pose a direct constitutional question. The case did, however, provide additional insight into the informational privacy issue and, in a dissenting opinion by Justice Brennan, reflected a growing concern among some members of the Court about the special dangers new information technology posed to privacy.

The *Miller* case began when the government sought banking records related to a criminal investigation. The bank, which had not been served a search warrant, voluntarily provided the records. The subject of the records challenged the government, maintaining he had been deprived of legal due process because the records were obtained by the government without a search warrant. To resolve the issue, the Court applied a "reasonable expectation of privacy" standard articulated by the Court in the 1967 case *Katz v. United States*.[10] In *Katz*, the Court held that the Fourth Amendment protected individuals from warrantless wiretaps, even if calls were monitored in telephone booths, not their homes. The Court reasoned that the Fourth Amendment protects individuals, not places, and that in American society people enjoy a "reasonable expectation of privacy" in places outside the home. The *Katz* holding also suggested that constitutional protections extend not only to personal property, but also to personal communications and information contained in them.

The *Miller* Court, however, reasoned that when individuals voluntarily turn records over to banks, the records become the property of the banks; consequently, the individuals relinquish any reasonable expectation of privacy with respect to those records.[11] Although the case was decided against the person on whom records were kept, it did reflect a nagging concern among some members of the Court about privacy problems that lay ahead because of computers and the proliferation of personal information in American society. In dissent, Justice Brennan said:

> A bank customer's reasonable expectation is that, absent a compulsion by legal process, the matters he reveals to the bank will be utilized by the bank only for internal banking purposes, affairs, opinions, habits, and association. Indeed, the totality of bank records provides a virtual current biography. Developments of photocopying machines, electronic computers, and other sophisticated instruments have accelerated the ability of government to intrude into areas which a person normally chooses to exclude from prying eyes and inquisitive minds. Consequently, judicial interpretations of the constitutional protection must keep pace with the perils created by these new devices.[12]

Whalen v. Roe—Trust in Government Computers

A year after *Miller*, the Court for the first time considered a case that dealt squarely with computerized records maintained by government. In *Whalen v. Roe*, the concerns about technology and personal privacy expressed in Justice Brennan's dissent in *Miller* were embraced by a majority on the Court.[13] The opinion in *Whalen* nudged the concept of informational privacy near, if not into, the constitutional realm.

In *Whalen*, several physicians and patients challenged the constitutionality of a New York statute that required doctors to provide the state with copies of prescriptions for certain kinds of controlled drugs, to be maintained in a state computer system. It was enacted as the Controlled Substance Act of 1972 by the New York Legislature in response to concerns that controlled drugs were being diverted into unlawful channels. The doctors argued that the drug-monitoring system interfered with their ability to practice medicine free from government interference; the patients argued the statute discouraged them from seeking needed medications. Both assertions were based on concerns that the state's computerized record system might be abused if improperly administered, resulting in violations of privacy. A federal district court ruled the New York statute was an unconstitutional violation of protected rights of privacy and enjoined its enforcement. The state appealed, and the U.S. Supreme Court agreed to hear the case.

The Court reversed the lower court and said the statute was a reasonable exercise of the state's broad police powers, that the plaintiff's concerns about the security of the state's computer system were speculative and not supported by facts presented in the case.[14]

Justice Stevens, writing for the seven-member majority, posed the legal issue before the Court in constitutional terms. He said the "constitutional question presented is whether the State of New York may record, in a centralized computer file, the names and addresses of all persons who have obtained, pursuant to a doctor's prescription, certain drugs for which there is both a lawful and unlawful market."[15]

In his analysis of the case, Justice Stevens recognized two kinds of constitutional privacy interests asserted by the patients and doctors. The first was informational privacy, or the interest of individuals in avoiding disclosure of personal matters. The second interest was decisional privacy, or independence in making certain kinds of personal decisions.[16] Although Justice Stevens recognized two such interests, he said insufficient evidence was presented to suggest the New York program violated either. With respect to the informational privacy claim, he noted that public disclosure of patient information could occur in three ways. Health department employees might violate the statute, either deliberately or negligently, that required them to maintain proper security; a patient or doctor might be accused of a crime and the computerized data taken as evidence in a criminal proceeding; or a doctor, pharmacist, or patient might voluntarily make the information public. He concluded the third possibility existed under prior law and was unrelated to the computer-filing program, and that neither of the first two concerns was sufficient to render the statute invalid. Regarding the computer security issue, Justice Stevens noted there was no support in the record, or in the experience of two states with similar statutes, to suggest New York's program would be administered improperly.[17]

Stevens and Brennan: Words of Caution

While the Court held that New York's computer-based program for monitoring controlled drugs did not violate a constitutional right of people to control information about themselves, the opinion expressed concerns about the impact of computers. Justice Stevens offered a "…final word about issues we have not decided" in the opinion, which echoed many reservations expressed in the legislative debates of the previous decade about computer privacy.[18] He also provided a constitutional footing for those concerns. Addressing what he viewed as potential constitutional pitfalls as society grew dependent on computers, he observed:

> We are not unaware of the threat to privacy implicit in the accumulation of vast amounts of personal information in computerized data banks or other massive government files. The collection of taxes, the distribution of welfare and social security and the enforcement of the criminal laws all require the orderly preservation of great quantities of information, much of which is personal in character and potentially embarrassing or harmful if disclosed. The right to collect and use such data for public purposes is typically accompanied by a concomitant statutory or regulatory duty to avoid unwarranted disclosure. Recognizing that in some circumstances that duty arguably has its roots in the Constitution, nevertheless New York's statutory scheme, and its implementing administrative procedures, evidence a proper concern with, and protection of, the individual's interest in privacy. We therefore need not and do not decide any question which might be presented by the unwarranted disclosure of accumulated private data whether intentional or unintentional or by a system that did not contain comparable security provisions. We simply hold that this record does not establish an invasion of any right or liberty protected by the Fourteenth Amendment.[19]

The ruling did provide some guidance about how the Court might deal with computer privacy and access conflicts in future cases. While the Court recognized the privacy interests of individuals in personal information, it also recognized that reasonable computer security precautions by administrative agencies were sufficient to ensure privacy. In other words, properly administered computerized record systems could be made as secure from unauthorized access as paper record systems. Taking this reasoning a step further, the opinion rejected the notion that the mere fact records were held in computers could justify blanket denial of public access to all records in the computer out of concern for privacy.

Justice Brennan, concurring in the result in *Whalen*, also expressed concerns about the effects of computer technology on personal liberties. He conceded that government had a legitimate interest in the collection and storage of personal data in computers, and that simply because new technology made government more efficient was not reason enough to render such activities unconstitutional. But he cautioned that broad dissemination of such private information would "clearly implicate constitutionally protected privacy rights…"[20] In words that anticipated Court actions, Justice Brennan said:

> …the Constitution put limits not only on the type of information the State may gather, but also on the means it may use to gather it. The central storage and easy accessibility of computerized data vastly increases the potential for abuse of the information, and I am not prepared to say that future developments will not demonstrate the necessity of some curbs on such technology.[21]

A dozen years after *Whalen*, developments in government use of computer technology brought the issue squarely back before the Court in *U.S. Department of Justice v. Reporters Committee for Freedom of the Press*, a case that could significantly reduce public access to government information held in centralized computers.[22]

REPORTERS COMMITTEE AND PRACTICAL OBSCURITY

In 1989, the Supreme Court decided a case that had lingered in the federal courts for more than a decade. In *U.S. Department of Justice v. Reporters Committee*, the Court arguably recognized a constitutional right of informational privacy on a par with other privacy interests acknowledged by the Court.[23]

Threats from Compiled Data

The case began its decade-long journey through the federal courts in 1978 when the Federal Bureau of Investigation denied a Freedom of Information Act (FOIA) request by CBS news correspondent Robert Schakne and the Reporters Committee for Freedom of the Press. The request was for "rap sheet" information—arrest and conviction data compiled in an FBI computer from various federal, state, and local law enforcement agencies. The information was about three brothers identified by the Pennsylvania Crime Commission as principals in a legitimate business dominated by organized crime that had dealings with a corrupt congressman. Ultimately, the request sought only public record information compiled by the FBI, which the D.C. Circuit of the U.S. Court of Appeals held should be released because public record information had little or no privacy interest.

The unanimous opinion by Justice Stevens in *Reporters Committee* has important implications for access to computerized records on several fronts. The paper focuses on the "practical obscurity" doctrine, a judicial acceptance of, among other things, the adage "forgive and forget." The Court's practical obscurity doctrine assumes an individual's right to control personal information and that computers exacerbate the threat to personal privacy by elimination of the natural elements of time and distance among "scattered bits of information" that once afforded individuals the ability to distance themselves from past mistakes and start their lives anew.

The doctrine also assumes that scattered bits of information about a person, when pieced together in the computer, create a composite that is more threatening than any separate bit of information. As a result, the court reasoned that information—even information compiled from scattered public records—gains a revitalized privacy interest when pooled in a centralized computer system.[24]

Justice Stevens, in his discussion of informational privacy and practical obscurity, appeared to adopt reasoning from Judge Star's dissenting opinion in the D.C. Circuit of the Court of Appeals, which turned on the fact that records were contained in government computers. Said Judge Star: "Computerized data banks of the sort involved here present issues considerably more difficult than, and certainly very different from, a case involving source records themselves."[25] Justice Stevens then set about to answer this question: Does the "compilation of otherwise hard-to-obtain information alter the privacy interests" in that information?[26] In so doing, he shifted the emphasis away from the public-record status of information and focused on the difficulty of obtaining the information and whether it was part of a compilation in a government database.

Stevens: Right to Control Information About Oneself

While *Reporters Committee* involved the statutory interpretation of an FOIA exemption, Justice Stevens posed the general question before the Court in broader, Constitutional terms. Citing his earlier opinion in *Whalen v. Roe*, Justice Stevens noted that privacy cases have asserted two constitutional values; individuals' interests in avoiding disclosure of personal information and independence in making certain kinds of intimate decisions. "Here, the former interest 'in avoiding disclosure of personal matters' is implicated," he said.[27]

Justice Stevens adopted a definition of informational privacy that "encompasses the individual's control of information concerning his or her person."[28] He promptly rejected as a "cramped notion of personal privacy" the assertion that criminal-record information had no privacy interest because it had previously been disclosed publicly.[29] Ignoring the fact that individuals generally have no control over disclosure of criminal records, he noted that both the common law and the literal understanding of privacy acknowledge the right of individuals to control information about themselves. To bolster this reasoning, Justice Stevens cited Webster's Dictionary[30] as well as legal commentators, including Alan Westin, whose book *Privacy and Freedom* and congressional testimony influenced the public debate over privacy and the computerization of government records. As Westin defined it: "Privacy is the claim of individuals... to determine for themselves when, how, and to what extent information about them is communicated to others."[31] Justice Stevens also cited the seminal 1890 article in the *Harvard Law Review* by Samuel Warren and Louis Brandeis, who later sat on the Supreme Court, that said: "The common law secures to each individual the right of determining, ordinarily, to what extent his thoughts, sentiments, and emotions shall be communicated to others...[E]ven if he has chosen to give them expression, he generally retains the power to fix the limits of the publicity which shall be given them."[32]

New Definition of 'Public Records'

After adopting a definition of informational privacy that encompassed at least some publicly available government records, Justice Stevens set about to distinguish between public and private information. He offered a definition of public records that turned not on the nature of the records, but on the nature of the system in which the records were kept and the difficulty of obtaining them. He reasoned that if the records sought by CBS and the Reporters Committee were truly "public," an FOIA request would not have been necessary. Said Justice Stevens:

> The very fact that federal funds have been spent to prepare, index, and maintain these criminal-history files demonstrates that the individual items of information in the summaries would not otherwise be "freely available" either to the officials who have access to the underlying files or to the general public. Indeed, if the summaries were "freely available," there would be no reason to invoke the FOIA to obtain information they contain. Granted, in many contexts the fact that information is not freely available is no reason to exempt that information from a statute generally requiring its dissemination. But hard-to-obtain information alters the privacy interest implicated by disclosure of that information. Plainly, there is vast difference between the public records that might be found after a diligent search of courthouse files, county archives, and local police stations throughout the country and a computerized summary located in a single clearinghouse of information.[33]

Justice Stevens cited several statutes and regulations to support the Court's conclusion that records that were public somewhere or at some time in government records systems could nonetheless be shielded from disclosure when compiled in government computers. He noted that Congress limited criminal record dissemination to banks, the securities industry, the nuclear power industry, law enforcement agencies, and local licensing agencies.[34] He also cited an FBI program that disseminated criminal-history records to law enforcement agencies but that threatened to terminate the sharing if the information were disclosed "outside the receiving departments or related agencies."[35]

To further support the Court's position, Justice Stevens cited the Privacy Act of 1974, which recognized the impact of computers on privacy. He said the Privacy Act could not be used to withhold information required to be disclosed under the FOIA, a conclusion Congress itself reached in 1984.[36] But he then ignored Congress' specific prohibition against using the Privacy Act to withhold public-record information and surmised that "Congress' basic policy concern regarding the implications of computerized data banks for personal privacy is certainly relevant in our consideration of privacy interests affected by dissemination of rap sheets from the FBI computer."[37]

Privacy for Those Identified in Records

Justice Stevens cited two examples in the FOIA itself to bolster his conclusion that Congress had not intended the disclosure statute to cover records of private citizens identifiable by name. He pointed out that the FOIA specifically provides that "[to] the extent required to prevent a clearly unwarranted invasion of personal privacy, an agency may delete identifying details when it makes available or publishes an opinion, statement of policy, interpretation, or staff manual or instructions."[38] He added that the FOIA required that "[any] reasonable segregable portion of a record shall be provided… after deletion of the portions which are exempt."[39] This requirement, he reasoned, was an acknowledgment by Congress that disclosure of records containing personal information about private citizens could infringe on significant privacy interests.[40]

In an additional attempt to distinguish between "scattered bits of criminal history and a federal compilation,"[41] Justice Stevens referred to the Court's opinion in *Department of Air Force v. Rose*, a case that recognized privacy interests in cadet disciplinary reports that at one time had been posted on bulletin boards at the United States Air Force Academy, a military educational institution not routinely open to the general public.[42]

In *Rose*, New York University law students writing about military discipline sought summaries of Air Force Academy Honor and Ethics Code violations. The law students sought only summaries with any identifying information deleted. The *Rose* Court held, however, that what constitutes identifying information must be weighted not only from the viewpoint of the general public, but also from the perspective of other cadets and academy personnel who might recognize the subjects of the disciplinary actions. Justice Stevens, equating public criminal records to institutional disciplinary records, seemed to suggest that records could endanger privacy and be withheld based on a very limited, speculative threat. He noted: "If a cadet has a privacy interest in past discipline that was once public but may have been 'wholly forgotten,' the ordinary citizen surely has a similar interest in the aspects of his or her criminal history that may have been wholly forgotten."[43]

Citing his earlier opinion in *Whalen*, Justice Stevens made clear in *Reporters Committee* that "wholly forgotten" criminal records regained privacy interests when compiled by government computers: "We are not unaware of the threat to privacy implicit in the accumulation of vast amounts of personal information in computerized data banks…"[44]

CONCLUSION

The Court's concerns about the impact of computers on personal privacy certainly were appropriate; when government gathers and stores personal information, it has a concomitant duty to ensure the privacy of that information. This is the essence of the Privacy Act of 1974 and subsequent legislation, and was the conclusion reached by the Court before *Reporters Committee*. Indeed, the rampant growth of non-governmental information vendors—Twitter, Facebook, Amazon, credit-card companies, big-data collectors, etc.—and their potential dissemination of personal data pose a serious threat to individual autonomy without the constitutional restraints imposed on a government information enterprise.

However, it is dangerous to equate personal information gathered by the government in the course of business with information relinquished by individuals that becomes part of a public record, be it at a local courthouse or some federal bureaucracy. The scope of public records has been defined over time, both by the common law and legislatures. These definitions are based on the nature and content of the information and the societal importance of public access to such information.

As the D.C. Circuit has noted, maintaining the obscurity of personal public-record information compiled in government databases is "attractive as a legislative policy matter" but not related to the statutory definition of privacy, which should be the concern of the courts.

The question of whether public-record information should enjoy a rejuvenated privacy interest is best left answered by the legislative process, which can systematically reassess the new dynamics of privacy in the computer age. As things stand after *Reporters Committee*, public and media access to public information might be determined not by the nature and content of the information, but rather by its form as part of a computer database or ease of access. This reasoning opens a loophole in public access legislation through which recordkeepers could shield information simply by including it in a computer compilation. Indeed, as computers increasingly dominate government record systems, scattered-about public records are becoming a thing of the past.

NOTES

1. *Sampson v. Murray*, 415 U.S. 61, 95 (1974).
2. John W. Macy Jr., "The New Computerized Age," *Saturday Review*, July 23, 1966, 15.
3. *Privacy Act*, 5 U.S.C. sec. 552a (1974). See also *The Computer and Invasion of Privacy: Hearings Before the Special Subcommittee on Invasion of Privacy of the House Committee on Government Operations*, 89th Cong., 2d Sess. (1966); and *Computer Privacy: Hearings Before the Subcommittee on Administrative Practices and Procedures of the Senate Committee on the Judiciary*, 90th Cong., 1st Sess. (1967).
4. *The Computer and Invasion of Privacy*, 6.
5. *Whalen v. Roe*, 429 U.S. 589, 606–607 (1977).
6. *U.S. Department of Justice v. Reporters Committee for Freedom of the Press*, 489 U.S. 749, 780 (1989).
7. *Paul v. Davis*, 424 U.S. 496 (1976).
8. *Paul v. Davis*, 713.
9. *United States v. Miller*, 425 U.S. 346 (1976).
10. *Katz v. United States*, 389 U.S. 347, 353 (1976).

11. *United States v. Miller*, 442–443.
12. *United States v. Miller*, 449, quoting *Burrows v. Superior Court*, 529 P.2d 590 (1974).
13. *Whalen v. Roe*, note 5.
14. *Whalen v. Roe*, 598.
15. *Whalen v. Roe*, 591.
16. *Whalen v. Roe*, 598.
17. *Whalen v. Roe*, 600.
18. *Whalen v. Roe*, 605.
19. *Whalen v. Roe*, 605–606. Justice Stevens cited Arthur Miller, "Computers, Data Banks and Individual Privacy; An Overview," *4 Colum. Human Rights L. Rev.* 1 (1972), and Arthur Miller, *The Assault on Privacy; Computers, Data Banks and Dossiers* (Ann Arbor, Mich.; University of Michigan Press, 1971). Both the law review article and book by legal scholar Miller, as well as his testimony before several congressional committees, were instrumental in passage of the Privacy Act of 1974.
20. *Whalen v. Roe*, 606.
21. *Whalen v. Roe*, 606–607.
22. *Reporters Committee v. DOJ*, note 6.
23. *Reporters Committee v. DOJ*, note 6.
24. *Reporters Committee v. DOJ* had other implications for access, which are not the focus of this paper. In addition to the practical obscurity doctrine and the recognition that individuals could control personal information about themselves, the Court held that agencies could determine that classes of information—in this case, compilations of public record information in computerized databases—could be "categorically" recognized as an unacceptable privacy threat and shielded from public disclosures without any meaningful balancing with competing societal interests, such as a public interest served by disclosure. The Court also reasoned that the sole purpose of disclosure under the FOIA was to shed light on the performance of agencies' statutory duties. This definition greatly narrows the scope of public interests that could overcome privacy concerns in access cases, and threatened to take out of the public realm vast amounts of information the government gathers and stores in computers that have no obvious bearing on agency performance.
25. *Reporters Committee v. DOJ*, 759–760.
26. *Reporters Committee v. DOJ*, 764.
27. *Reporters Committee v. DOJ*, 762.
28. *Reporters Committee v. DOJ*, 763.
29. *Reporters Committee v. DOJ*, 762.
30. *Reporters Committee v. DOJ*, 763–764.
31. *Reporters Committee v. DOJ*, 764.
32. *Reporters Committee v. DOJ*, 764.
33. *Reporters Committee v. DOJ*, 764.
34. *Reporters Committee v. DOJ*, 753.
35. *Reporters Committee v. DOJ*, 752, citing 28 U.S.C. sec. 534(b). Interestingly, the concern about widespread dissemination expressed in 28 U.S.C. sec. 534(b) was not about public record information contained in criminal history files, but rather for uncorroborated or hearsay information and outdated information, such as charges that were filed but later dropped. Also, the legislative history of the original FOIA Exemption 7(c) focused on the potential harm disclosure would pose to investigations, not to individuals.

36. *Reporters Committee v. DOJ*, 765, citing the Privacy Act of 1974 and quoting H.R. No. 93-1416, 7 (1974).
37. Congress, reacting to several conflicting court opinions that attempted to use the Privacy Act as justification for withholding information, amended the Privacy Act in 1974 to specifically state that the act was not an Exemption 3 statute under the Freedom of Information Act. The new language was achieved as an amendment to the National Security Act of 1947 (50 U.S.C. sec. 431). See Thomas M. Susman, "The Privacy Act and the Freedom of Information Act: Conflict and Resolution," *John Marshall L. Rev.* 21 (Summer 1988): 703.
38. *Reporters Committee v. DOJ*, 765.
39. *Reporters Committee v. DOJ*, 765.
40. *Reporters Committee v. DOJ*, 765. The FOIA language, however, also was an acknowledgment by Congress that all information held by agencies was disclosable, unless covered by a specific FOIA exemption, of which the Privacy Act was not. See note 37.
41. *Reporters Committee v. DOJ*, 767.
42. *Department of Air Force v. Rose*, 425 U.S. 352 (1976).
43. *Reporters Committee v. DOJ*, 769.
44. *Reporters Committee v. DOJ*, 770, citing *Whalen v. Roe*, 605.

CHAPTER TWO

Tipping the Scales: How the U.S. Supreme Court Eviscerated Freedom of Information in Favor of Privacy

MARTIN E. HALSTUK, BENJAMIN W. CRAMER, & MICHAEL D. TODD

At his inauguration on January 20, 2009, President Barack Obama invoked a vow he made repeatedly throughout his presidential campaign.[1] He had promised that, if elected, he would create a sweeping transparency policy.[2] The day after his inauguration, President Obama issued three directives—one executive order[3] and two presidential memoranda[4]—instructing the federal bureaucracy to heighten disclosure under the Freedom of Information Act.[5] "Transparency and rule of law will be the touchstones of this presidency. ... [T]his administration stands on the side not of those who seek to withhold information, but those who seek to make it known," President Obama said.[6]

President Obama's first acts were symbolic of his vow to run a clean and open government. Were his promises fulfilled? With advances in technology and e-government, just how well is the U.S. federal government doing in providing people the information it says it should provide, and what role does privacy play in denying citizens access to their records?

To get at these questions, this chapter will build on Chapter 1 by laying out the basics of the U.S. Freedom of Information Act and describing in more detail how the Supreme Court has interpreted the Act in relation to personal privacy, essentially weakening the law to provide greater protections for privacy.[7] Despite these court rulings eviscerating FOIA, President Obama said his agencies will be more open than preceding administrations. To test that, the authors compare FOIA compliance between Obama and former President George W. Bush. Did increased transparency really occur, as Obama promised? The findings might be surprising to some.

Ultimately, at the end of this chapter we conclude that the High Court has reformulated the congressionally intended framework to "pierce the veil" of government secrecy and promote government accountability.[8] The result is a substantial reduction in the categories of government-held information available to the public, regardless of what presidents say they will do. The Court, we argue, in executing its thesis of privacy exceptionalism, has overreached its judicial authority, taking upon itself the role of the ultimate arbiter of the meanings of the FOIA's two privacy exemptions, regardless of congressional intent or presidential FOIA policy, be it the policy of President Bush, President Obama or future presidents.

THE FOIA

The U.S. Freedom of Information Act, which has been amended in significant respects since its enactment in 1966, grants the public a statutory right to examine and obtain copies of the mountains of paper and photos and billions of bytes of digitally formatted information collected by federal agencies and departments.[9] The FOIA's disclosure mandate is limited only by the Act's statutory exemptions created by Congress.[10]

Journalists have used the FOIA to expose waste and corruption in government and to reveal unsafe consumer products, harmful drugs, and public health hazards. Historians have gleaned important information to help them fill crucial gaps in history.[11]

The FOIA's extensive legislative history articulates Congress' view that the statute advances America's bedrock democratic principles in many ways. For example, transparency holds government officials and other decision makers publicly accountable for their actions, openness enables the general public to make informed decisions at the polls, and full disclosure helps the press to achieve its constitutionally designated role as a government watchdog.[12]

Over the decades, Congress has repeatedly stressed that a presumption of government openness undergirds the FOIA.[13] During the FOIA's first few years, agencies virtually ignored the law.[14] Agency bureaucrats used various ploys to discourage FOIA use, such as charging exorbitant fees for copying documents, delaying responses and claiming they could not find the requested records.[15]

Congress responded sharply in its first amendment to the law, passed in 1974. The House excoriated agency disregard for the FOIA, asserting that operation of the Freedom of Information Act has been "hindered by five years of foot-dragging," and "widespread reluctance of the bureaucracy to honor the public's legal right to know."[16] The 1974 amendments clarified poorly crafted or vague language contained in the original 1966 statute.[17]

In 1976, Congress again amended the FOIA, specifically to counter a Supreme Court decision in which the Court broadly interpreted agency discretion to refuse disclosure.[18] Lawmakers added new language to explicitly limit agency discretion to withhold information, and to reiterate the statute's strong presumption of openness.[19]

In the late 1980s, by which time the government had computerized most of its information gathering, agencies routinely rejected FOIA requests for information contained in electronic formats and databases, arguing that neither the Act's language nor legislative history required disclosure in computerized formats. In response, Congress passed the Electronic Freedom of Information Act (EFOIA) of 1996.[20] This amendment required agencies to release records regardless of the form or format in which they were contained. Further, agencies were also directed to provide the information in any form or format that FOIA requesters wished.[21]

PROTECTIONS FOR PERSONAL PRIVACY

Congress' actions over the decades helped bolster the public's right to know as embodied in the FOIA. However, the Act's original lawmakers also understood fully that the public interest in government-held information is not absolute.[22] They realized that citizens in a participatory democracy must have access to government information,[23] but under certain limited circumstances confidentiality of government-held information is necessary.[24] Hence, the FOIA's creators included nine statutory exemptions under which agencies may withhold information.[25]

Specifically, in regard to the privacy protections evinced in Exemptions 6 and 7(C), the FOIA's legislative history and Supreme Court decisions have recognized that the right of privacy and the public's right to information represent two equally vital societal values. Congress made it abundantly clear that in order to resolve the tension between these rights it is not necessary that "either be abrogated or substantially subordinated."[26] Through the privacy exemptions, lawmakers conveyed their understanding that, under specifically enumerated exceptions, a person's right to privacy may trump society's democratic need to find out what the people's government is doing.[27]

To resolve this tension, the law must maintain an appropriate balance between a person's privacy and society's need to be informed. As the noted privacy scholar Alan F. Westin observed in his seminal 1967 work, *Privacy and Freedom*, democracies must "set a balance between government's organizational needs for preparatory and institutional privacy and the need of the press, interest groups, and other governmental agencies for the knowledge of government operations required to keep government conduct responsible."[28]

It is precisely such a balance that Congress sought to create with personal privacy Exemption 6 and law enforcement-related privacy Exemption 7(C). Exemption 6 states that agencies may withhold "personnel and medical files and similar files the disclosure of which would constitute a clearly unwarranted invasion of personal privacy."[29] To determine whether a FOIA request triggers Exemption 6, agencies must (1) determine if the information falls within the definition of "personnel," "medical," or "similar" files, and (2) then balance the privacy interests against the public's interest in disclosure to determine whether the release of the requested information would be clearly unwarranted.[30]

Under Exemption 7(C), investigatory "records or information compiled for law enforcement purposes" may be withheld.[31] Exemption 7(C) was designed to protect private identifying information that could endanger law enforcement personnel, their families, and confidential informants.[32] Under Exemption 7(C), agencies can withhold law enforcement information which, if released, "could reasonably be expected to constitute an unwarranted invasion of personal privacy."[33] Exemption 6 requires the government to show that disclosure "would constitute a clearly unwarranted invasion of privacy,"[34] whereas the law enforcement exemption provides agencies with a lower privacy standard to substantiate withholding a requested record.[35] Courts have since held that law Exemption 7(C) allows law enforcement agencies greater latitude to refuse disclosure in order to protect privacy than under the stricter Exemption 6 standard.[36]

BALANCING TEST

The solution to settling the conflict between individual privacy and government transparency was perhaps most clearly articulated by Justice William J. Brennan, who wrote the majority opinion in *Department of the Air Force v. Rose*,[37] the first FOIA-related privacy lawsuit to reach the Supreme Court.

In this 1976 opinion, which concerned Exemption 6, a 5-to-3 majority held that when the government uses the personal privacy exemption to withhold a record, courts must use a balancing test to determine whether a requested record should be disclosed.[38]

The Department of the Air Force triggered Exemption 6 to withhold records concerning cadet cheating at the U.S. Air Force Academy, arguing that disclosing such records would stigmatize the cadets for the rest of their careers.[39]

Justice Brennan, writing for the majority, reasoned that the release of disciplinary actions with the names of the individual cadets deleted was in the public interest.[40] Brennan wrote that the strong public interest in cheating and other violations of discipline at the academy was obvious because obedience and reliability are important to military effectiveness.[41]

In upholding the ruling by the U.S. Court of Appeals for the Second Circuit, which ordered the Air Force to release the summaries, Brennan said the Act's legislative history makes it "crystal clear" that the congressional objective for the FOIA was to allow the public to evaluate the government's performance and promote governmental accountability.[42]

In effect, other considerations such as the potential embarrassment to the Air Force Academy should not be part of the equation.[43] He emphasized further that nondisclosure under FOIA's exemptions in general apply only to limited categories of information enumerated in the FOIA, and the Act's statutory language and legislative history indicate that the exemptions must be narrowly construed. These "limited exemptions do not obscure the basic policy that disclosure, not secrecy, is the dominant objective of the act."[44]

In the first Exemption 7(C) case decided by the Supreme Court in *Federal Bureau of Investigation v. Abramson*, a bare 5-to-4 majority in 1982 upheld an agency decision to reject a journalist's FOIA request made to the FBI for documents concerning individuals who had criticized the presidential administration.[45] Justice Day O'Connor, one of the four Abramson dissenters who supported disclosure, reproached the Court majority for its decision.

> It scarcely needs to be repeated that Congress' ultimate objective in requiring such disclosure was to 'ensure an informed citizenry, vital to the functioning of a democratic society, needed to check against corruption and to hold the governors accountable to the governed.'[46]

Echoing the *Rose* Court's majority opinion by Brennan, Justice Day O'Connor strongly suggested that the *Abramson* majority contravened the FOIA's legislative history and Congress' intent. Congress is "free to draw lines [between exempt and nonexempt materials] without cavil from this Court, so long as [Congress] respects the constitutional proprieties," she said.[47]

Including *Rose* and *Abramson*, the Supreme Court has heard eight privacy-related FOIA disputes in all. With the exception of *Rose*, the Supreme Court sided with the government every time.[48] This analysis turns now to four of these Supreme Court decisions. The authors selected these four opinions because, in the aggregate, they represent a body of FOIA jurisprudence that seems to contradict the view that "disclosure, not secrecy, is the dominant objective of the act."[49]

ESTABLISHING THE DOCTRINE OF PRIVACY EXCEPTIONALISM

The trend to diminish the FOIA-related public interest began in 1982, when the Supreme Court overruled the U.S. Circuit Court of Appeals for the District of Columbia, and held that even a minimal privacy interest—one that touches on non-intimate information—is sufficient to raise Exemption 6 as a bar to disclosure.[50]

In *Department of State v. Washington Post*, reporters made a FOIA request to confirm a report that two officials of Iran's revolutionary anti-American government were traveling under U.S. passports.[51] The State Department denied the request, citing Exemption 6's protection for "similar files" that trigger privacy interests.[52]

The D.C. court rejected the government's argument and ordered the State Department to release the information to the newspaper reporters. The appeals court reasoned that the privacy interests in the citizenship status of the two Iranians was significantly less intimate than information normally contained in "personnel, medical and similar files" as required in Exemption 6.[53]

In overturning the D.C. appeals court ruling, Justice William Rehnquist, writing for the Court, contended that the passport was subject to withholding under Exemption 6. In his analysis, Rehnquist said withholding the passport information was supported by the *Rose* opinion, which held that Exemption 6 does not exempt every incidental invasion of privacy—it protects only those disclosures that would constitute clearly unwarranted invasions of personal privacy.[54]

Rehnquist, one of the dissenters in the pro-disclosure *Rose* case, reasoned that the question before the *Washington Post* Court turned on the meaning of an "incidental" invasion of privacy.[55] In Rehnquist's view, disclosure of the passport information "would constitute a clearly unwarranted invasion of personal privacy." Furthermore, Rehnquist held that even a minimal privacy interest is sufficient to trigger an Exemption 6 analysis.[56]

According to Rehnquist's interpretation of the FOIA's legislative history, Congress intended for Exemption 6 "to protect individuals from the injury and embarrassment that can result from the unnecessary disclosure of personal information."[57] Rehnquist conceded that personal identifying information, such as a person's place of birth, date of birth, date of marriage, employment history and comparable data are "not normally regarded as highly personal."[58]

Nonetheless, such non-intimate information may be withheld if disclosure would constitute a clearly unwarranted invasion of personal privacy. "In sum, we do not think that Congress meant to limit Exemption 6 to a narrow class of files containing only a discrete kind of personal information." In his view, the citizenship information requested by the reporters satisfies the "similar files" requirement of Exemption 6.[59]

Although the *Washington Post* opinion appeared to be based on an arguably revisionist interpretation of Exemption 6's legislative history, the holding nonetheless broadened Exemption 6's scope and diminished the FOIA related public interest.[60]

In 1989, the Court handed down a second anti-disclosure ruling that substantially expanded the boundaries of the FOIA's privacy interests. This Exemption 7(C) case concerned a journalist's FOIA request to aid in his investigation into alleged political corruption involving a congressman and a mob-linked business that received government defense contracts.

THE 'CENTRAL PURPOSE' TEST

In *Department of Justice v. Reporters Committee for Freedom of the Press*, the Court upheld the government's decision to reject the late CBS reporter Robert Schakne's FOIA request for an FBI computerized "rap sheet" of Pennsylvania businessman Charles Medico. Medico, who was alleged by the Pennsylvania Crime Commission to have ties to organized crime, was one of the owners of Medico Industries, a legitimate Pennsylvania company that received defense contracts.[61]

Schakne requested Medico's criminal history record to aid in his investigation into Charles Medico's relationship with Daniel J. Flood, who was a Pennsylvania congressman. After the FBI denied Schakne's request, Schakne and the Reporters Committee for Freedom of the Press sued to obtain the records.[62]

The Supreme Court reversed the decision by the U.S. Circuit Court of Appeals for the District of Columbia, which ordered the FBI to release Medico's rap sheet. In reaching its decision, a key part of the High Court's rationale was its definition of the FOIA's "central purpose." The Court, then headed by Rehnquist who succeeded former Chief Justice Burger, asserted that Congress' intent for the FOIA was to enable the public to evaluate government operations and performance.

Justice John Paul Stevens, writing for the Court, held that the Act's "central purpose" was "to ensure that the government's activities be opened to the sharp eye of public scrutiny, not that information about private citizens that happens to be in the warehouse of the government be so disclosed."[63] In other words, the Act applies only to official information that directly sheds light on the performance or conduct of a governmental agency or official.[64]

Stevens reasoned that the rap sheets "would tell us nothing directly about the character of [U.S. Rep. Flood's] behavior. Nor would it tell us anything about the conduct of the Department of Defense in awarding one or more contracts to the Medico Company."[65] Therefore, Stevens concluded, Medico's criminal history records fall "outside the ambit of the public interest that the FOIA was enacted to serve,"[66] and their disclosure could constitute an unwarranted invasion of privacy under Exemption 7(C).

Although the Court unanimously agreed that the government may deny Schakne's FOIA request, Justice Blackmun wrote a concurring opinion arguing that the majority position was overbroad. Blackmun, who was joined by Brennan, said that Stevens' opinion exempting "all" rap-sheet information from the FOIA's disclosure requirements conflicted with Exemption 7(C)'s plain language, its legislative history, and prior case law.⁶⁷

In its majority holding, the Rehnquist Court fortified a now clearly emerging FOIA privacy exceptionalism thesis with its reformulation of the Act's "central purpose." Two years later, the Court continued to restrict the ambit of information subject to disclosure under the FOIA when justices declared that the Court accords a "presumption of legitimacy" to government evaluations of government performance and honesty.⁶⁸

'PRESUMPTION OF LEGITIMACY' RATIONALE

In *Department of State v. Ray*, a 1991 personal privacy Exemption 6 lawsuit, the Court declared that the judiciary generally accords "government records and official conduct" a presumption of legitimacy.⁶⁹

In reaching its decision, the *Ray* Court overturned a ruling by the U.S. Court of Appeals for the Eleventh Circuit, which ordered the government to supply an immigration attorney with contact information for undocumented Haitian immigrants who unsuccessfully sought asylum in the United States.⁷⁰

The State Department rejected the lawyer's request, relying on Exemption 6. The lawyer sued in the U.S. District Court for the Southern District of Florida to obtain the contact information of the Haitians who were deported.

The U.S. government flatly rejected the attorney's fears of persecution of his former clients as unfounded because, the State Department maintained, American representatives had been monitoring the treatment of deported Haitians and said they were satisfied that the Haitian government made good on its assurances that it would not harass or prosecute the returnees. The State Department provided the immigration lawyer with 96 pages of documents that contained summaries of a number of interviews with specific returnees and the government's findings, but all the names were redacted.⁷¹

However, the district court ordered the State Department to release the missing information, reasoning that the public interest in the "safe relocation of returned Haitians" outweighed any invasion of the personal privacy arising from the "mere act of disclosure of names and addresses."⁷² The Eleventh Circuit affirmed, holding that the benefit of giving the attorney the means to locate his former clients outweighed the privacy interests of the deported Haitians.

The Supreme Court overturned the Eleventh Circuit and upheld the Department of State's decision to withhold the Haitians' identifying information.⁷³ Citing the *Reporters Committee* definition of the FOIA's "central purpose" test, Justice John Paul Stevens, writing for a unanimous Court, said that releasing the names of the refugees would not shed light on the official actions of the State Department.⁷⁴

Stevens said that the redacted summaries were sufficient for confirmation that no harm came to the returnees.[75] "Mere speculation about hypothetical public benefits" arising from efforts to confirm the veracity of the U.S. government's assertions" cannot outweigh a demonstrably significant invasion of privacy."[76]

Second, in expounding on the State Department's disclosure obligations under the FOIA in general, Stevens said, "We generally accord government records and official conduct a presumption of legitimacy."[77] Stevens discounted the immigration lawyer's request as simply a "hope" that the redacted information could be used to discover information not available in the official summaries. There is "not a scintilla of evidence in either the documents themselves or the record that tends to impugn the integrity of the reports," he wrote.[78]

The *Ray* FOIA opinion significantly expanded the Court's privacy exceptionalism thesis to block government transparency in two significant ways. First, it reaffirmed the Court's commitment to the narrowly defined Court-created "central purpose" standard set forth in *Reporters Committee*. Second, it applied a Court-crafted "presumption of legitimacy" rationale, shifting the burden of proof to FOIA requesters to provide convincing evidence in order to obtain governmental information. This shifting of the burden of proof is contrary to the democratic principles that undergirds the FOIA, as evinced in the Act's exhaustive legislative history:[79] the government must carry the burden of proof under a presumption of disclosure.[80]

The fourth and final case in this chapter addresses yet another court-created test under which FOIA requesters carry a heavy burden of proof to justify acquiring government-held information.

HEAVY BURDEN OF PROOF FOR FOIA USERS

In *National Archives and Records Administration v. Favish*, a law enforcement Exemption 7(C) privacy case, the Supreme Court laid the final stone in its construction of the Doctrine of FOIA Privacy Exceptionalism.[81] The *Favish* Court held that if the purpose of a FOIA request is to investigate whether officials acted negligently or otherwise improperly in performance of their duties—and the government raises Exemption 7(C) as a bar to disclosure—the requester must then produce evidence of "misfeasance or another impropriety" in advance of disclosure in order to overcome the "presumption of legitimacy" accorded to official government conduct and records.[82]

The *Favish* case concerned a FOIA request for copies of all death scene and autopsy photos taken by government agencies that investigated the suicide of White House Deputy Counsel Vincent Foster, Jr.

Foster, a friend of President Clinton and Hillary Clinton, was found dead with a gun in his hand in a Virginia park just outside Washington, D.C., in July 1993.[83] The U.S. Park Police conducted the initial investigation, and officers took color photos of the death scene, including ten pictures of Foster's body. Park police investigators concluded that Foster took his own life by shooting himself.[84] Four subsequent independent investigations, by the Federal Bureau of Investigation, Congress, and two separate Offices of the Independent Counsel (OIC), concurred that Foster committed suicide.[85]

The death prompted a number of conspiracy theories. One such theorist, a skeptical Los Angeles attorney, wanted to conduct an independent investigation and requested all the photos taken at the death scene and afterward.[86] Eventually, the government released more than 100 photos but withheld several of the more graphic and gruesome pictures, arguing that disclosure could constitute an unwarranted invasion of privacy of Foster's family members under Exemption 7(C). The attorney sued to obtain the remaining photos, and the Court upheld the government's decision to withhold them. In so holding, the Supreme Court recognized for the first time that FOIA-related privacy interests apply to surviving family members of deceased subjects of a FOIA request.[87] The Court thus resolved the question of so-called FOIA relational-privacy rights (also known as survivor-privacy rights), a controversial and emotional issue that lower courts had grappled with for more than a quarter-century, typically holding in favor of disclosure.[88]

However, the Court went beyond settling the immediate question of FOIA-related relational privacy rights. In arriving at its conclusion, the Court created en route a strict and unprecedented test for disclosure under the FOIA, to be applied whenever a FOIA requester seeks information to investigate government malfeasance and an agency raises a privacy exemption to defend withholding.[89]

Justice Anthony Kennedy acknowledged that the FOIA embodies a presumption in favor of disclosure: When documents are requested under the FOIA, requesters are not required to give reasons for why they want the information, nor need they explain how they may use the information.[90]

Kennedy added, however, that when a FOIA request raises privacy interests protected under Exemption 7(C)—and the purpose of the request is to investigate either government malfeasance or whether "responsible officials acted negligently or otherwise improperly in performance of their duties"[91]—then the "usual rule that the citizen need not offer a reason for requesting the information must be inapplicable."[92] Kennedy said that under such circumstances a FOIA requester must meet a two-prong test to establish "a sufficient reason" for disclosure. First, under the "sufficient reason" test the requester must show that there is a significant public interest in the requested information. Second, the requester must demonstrate that disclosure is likely to advance that significant public interest.[93] Relying on the *Ray* Court holding, Kennedy added that in order to satisfy the second prong of the test, the requester must produce evidence of "misfeasance or another impropriety" in advance of the disclosure in order to overcome a "presumption of legitimacy" accorded to official government conduct and records.[94]

In applying the *Ray* Court's "presumption of legitimacy" doctrine to *Favish*, the Court reasoned that *Favish* had not provided any evidence that "would warrant a belief by a reasonable person that the alleged government impropriety might have occurred."[95] Hence, the Court granted the government's motion for summary judgment to withhold the remaining disputed photos.

PRESIDENTIAL APPLICATION OF FOIA

While the courts hash out the balance between access to public records and personal privacy, presidents might have their own views toward transparency. Congress made FOIA, the Supreme Court interprets it, and the executive administers it. Throughout the 2008 presidential race, Obama repeatedly asserted that the George W. Bush administration engaged in eight years of unprecedented government secrecy.[96] On taking office, President Obama released two memos and one Executive Order admonishing the executive branch agencies to comply with his directives to enforce federal agency compliance with the FOIA.[97]

In his Memorandum on Transparency and Open Government, the President declared that the government should be transparent, participatory and collaborative.[98] He reminded agencies to make federal agency records available in formats "the public can readily find and use,"[99] including digitally recorded information.[100]

The second directive, the Memorandum on the Freedom of Information Act, declares that the Obama Administration's goal is to become "measurably more pro-disclosure than any previous administration."[101] In the face of doubt over whether or not an agency should disclose a record, the Administration's policy is that "record openness prevails."[102]

In addition to the two memoranda, President Obama issued Executive Order 13,489,[103] which reversed President George W. Bush's controversial Executive Order 13,233 of 2001.[104] Bush's order restricted public access to the papers of past presidents, thereby overriding the 1978 Presidential Records Act (PRA).[105] Before the Bush Order, presidential papers were subject to disclosure under the FOIA twelve years after a president leaves office as per the PRA. President Obama's Executive Order restored control of the Presidential Papers to the PRA and FOIA's disclosure provisions.

Traditionally, the Department of Justice (DOJ) issues its official FOIA compliance policy whenever a new presidential administration takes office. In Attorney General Eric Holder's FOIA policy memorandum, issued March 19, 2009, Holder reiterated that the Obama Administration's FOIA disclosure philosophy shall be based upon a "presumption of openness."[106]

Holder emphasized that a FOIA request may be refused only if the material falls under one or more of the FOIA's exemptions. He noted that the statute's language clearly states that FOIA exemptions are discretionary—not mandatory. The Administration "strongly encourages agencies to make discretionary [decisions in favor of] disclosures," he said.[107] Holder further instructed agencies that when they choose not to disclose a requested record because some of it contains classified material or other exempt information, then the agencies are required under the FOIA "to take reasonable steps to segregate and release nonexempt information" contained in that record.[108] The question is, did agencies follow that presidential directive?

THE STUDY

In order to measure the effects of Obama's directives on government transparency, the authors conducted a study of federal agency FOIA compliance based on data collected by the Department of Justice. The Justice Department is the FOIA's statutorily designated office to oversee and enforce the Act's procedures.[109]

This research project relies on reports issued by seven departments and five agencies as a representative sample. These twelve offices were selected because the total of FOIA requests they receive annually represents an average of about two-thirds of all FOIA requests received government-wide.[110]

The study compared the first three fiscal years of the Obama presidency (FY 2009–2011—the most recent data available) with the last three fiscal years of the George W. Bush presidency (FY 2006–2008).[111] This study is limited to an examination of federal agency disclosure practices pertaining to the FOIA's two privacy exemptions. Except for the Department of Homeland Security, the two privacy exemptions account for the majority of all FOIA requests denied government-wide:[112] personal privacy Exemption 6 and law enforcement privacy Exemption 7(C).[113] According to the official U.S. Department of Justice, Freedom of Information Act Guide & Privacy Act Overview, privacy Exemption 6 is most often used to deny FOIA requests.[114] When one also includes Exemption 7(C), the exemption for law enforcement records that implicate personal privacy, the executive branch offices have a formidable privacy rationale to refuse FOIA requests.

The authors' methodology focused on three specific areas of data pertaining to the privacy exemptions: (1) the numbers of FOIA requests processed for each of the 12 offices during each of the six fiscal years examined; (2) the numbers and rates of FOIA requests rejected by each office under Exemption 6; and (3) the number and rates of denials under FOIA Exemption 7(C).

This study found that in comparing the two 3-year average rates of FOIA requests outcomes of the two presidential administrations, the Obama Administration's average rate for nondisclosure was slightly to moderately *higher* than the Bush

Administration's average in 16 out of 24 categories in instances when the government used Exemptions 6 or 7(C) to refuse a FOIA request. In other words, while Obama insisted on greater openness, these data indicated greater secrecy, particularly when the information requested involves some element of personal privacy.

3-YEAR-AVERAGES OF FOIA DENIALS BY BUSH AND OBAMA ADMINISTRATIONS, CITING PRIVACY EXEMPTIONS 6 AND 7(C): FY 2006 THROUGH FY 2011

Table I. Personal Privacy Exemption 6 Denials

Federal Offices	Bush: FY 2006–2008	Obama: FY 2009–2011
Agriculture	61%	63%*
Defense	65%	70%*
Homeland Security	62%	71%*
Justice	49%	56%*
State	45%*	30%
Treasury	25%	27%*
Labor	25%	26%*
EEOC	5%	12%*
NLRB	37%*	31%
NARA	46%	80%*
SEC	9%	22%*
EPA	20%	23%*

* Indicate the greater percentage of FOIA denials in the comparisons of each agency and each department in Table I. The Obama record exceeds the Bush record in 10 of 12 offices above.

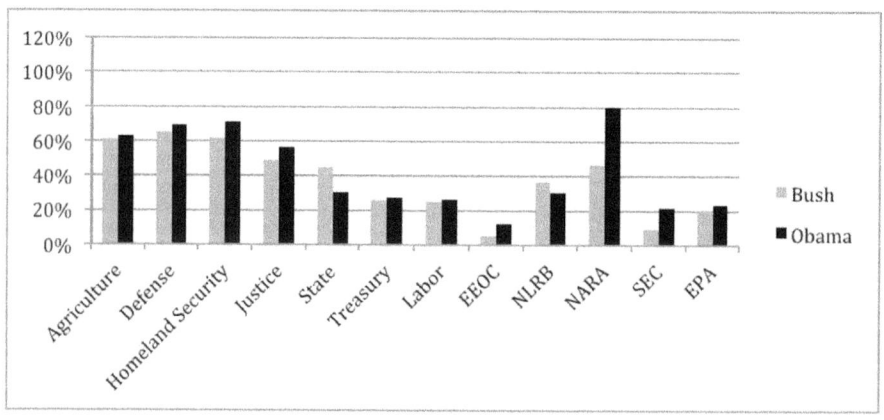

Fig 1. Bar chart representation for 3-year-averages of FOIA denials citing Exemption 6 above.

Table II. Law Enforcement Exemption 7(C) Denials

Federal Offices	Bush: FY 2006–2008	Obama: FY 2009–2011
Agriculture	31%*	28%
Defense	30%	35%*
Homeland Security	86%*	81%
Justice	63%	68%*
State	6%*	2%
Treasury	30%*	29%
Labor	57%	63%*
EEOC	18%	33%*
NLRB	39%*	34%
NARA	12%	23%*
SEC	5%	21%*
EPA	9%*	7%

* *Indicate the greater percentages of FOIA denials in the comparisons of each agency and each department in Table II. The Obama record exceeds the Bush record in 6 of 12 offices above.*

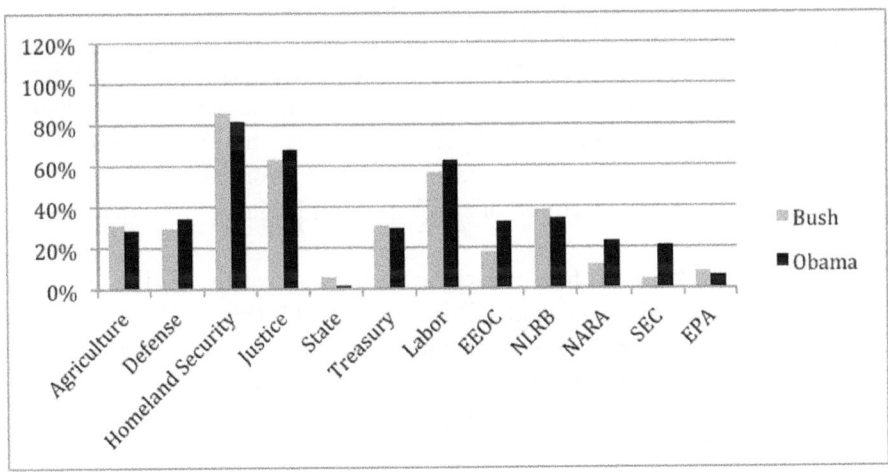

Fig 2. Bar chart representation for 3-year-averages of FOIA denials citing Exemption 7(C) above.

SUPREME COURT IS KING

On President Obama's first full day in office, he issued two official memoranda and one Executive Order intended to promote greater transparency under the FOIA. The public response was positive. The *Wall Street Journal*, acknowledging that the Bush administration openly directed federal offices subject to the FOIA to be cautious in releasing information, observed that the Obama Administration was clearly adopting a different philosophy of governance when it comes to openness. The *New York Times* trumpeted, the "promise of transparency is heartening."

Yet, three years later, this promise remained unfulfilled.

As this chapter sought to demonstrate, entrenched practices of withholding documents from the public and press continue unabated at federal agencies and departments. A comparison of FOIA request denials in the last three fiscal years of the Bush Administration and the first three fiscal years of Obama's term in office show virtually no change. Of particular concern is that the government's uses of the FOIA's two statutory privacy exceptions—personal privacy Exemption 6 and law enforcement privacy Exemption 7(C)—have exhibited a slight upward trend since Obama became president.

This failure to achieve greater government transparency through the FOIA may not be the fault of agency officials, nor may it be under the control of the Obama Administration, indeed, of any presidential administration, regardless of motivations for secrecy or wishes to run an open government.

In reality, the thesis advanced in this study asserts that the U.S. Supreme Court has preempted control over the FOIA and, in doing so, circumvented the constitutional scheme of checks and balances achieved through a tripartite separation of powers. Congress passed the FOIA. It is the statutory responsibility of the executive branch to enforce the Act. And it is the role of the judiciary to interpret the Act in those instances when a disclosure issue ends up in the courts for resolution.

But in practice, the Supreme Court has exploited the FOIA's two privacy exemptions to substantially reduce the categories of information subject to the FOIA's mandate, in contravention of the Act's plain language and extensive legislative history.

As demonstrated in the four privacy cases under study here, the U.S. Supreme Court has incrementally fashioned a Doctrine of FOIA Privacy Exceptionalism. The result is a court-created policy—tantamount to legislative action—to broadly construe the scope of the FOIA's privacy exemptions and thereby diminish the Act's public interest in disclosure, whenever the government raises privacy interests as a bar to disclosure.

This doctrine is grounded in five legal principles that the Court established in four cases: *National Archives and Records Administration v. Favish*, *U.S. Department of State v. Ray*, *Department of Justice v. Reporters Committee for Freedom of the Press*, and *Department of State v. Washington Post Co.*

Under this paradigm, an individual's privacy interests nearly automatically trump the societal public interest in disclosure. These four Court rulings represent a presumption of nondisclosure that stands in stark contrast to FOIA's voluminous legislative record. Nothing in the FOIA's legislative history indicates that Congress intended (1) that a minimal privacy interest—one that merely touches on non-intimate information—is sufficient to trigger a privacy exemption;[115] or (2) that the FOIA's "central purpose" is to disclose only information that directly sheds light on official government activities;[116] or (3) that Courts must accord a "presumption of legitimacy" to government reports;[117] or (4) that FOIA requesters must demonstrate that disclosure would advance a substantial public interest;[118] or (5) that FOIA requesters investigating allegations of government wrongdoing must offer evidence of wrongdoing in advance in order to get the information.[119]

Prior to the Rehnquist Court's privacy exceptionalism thesis, courts historically had balanced the public interest in disclosure against the individual's privacy interests when considering FOIA requests for information. The legislative history states, and the courts have traditionally held, that privacy interests do not necessarily outweigh the public interest in disclosure, and the government must carry the burden of justifying nondisclosure.

Under the Court's Doctrine of FOIA Privacy Exceptionalism, however, the threshold for a successful privacy exemption challenge is so low that the FOIA has been turned on its head by a court-crafted presumption of nondisclosure. It is precisely this Court-created doctrine that has obstructed President Obama's efforts to achieve a truly transformative change in the direction of greater government openness. In the final analysis, the Supreme Court's Doctrine of FOIA Privacy Exceptionalism is an example of judicial overreaching that is in direct conflict with the bedrock democratic principles of transparent governance and accountability of public officials for their actions.

NOTES

1. Sheryl Gay Stolberg, "On First Day, Obama Quickly Sets a New Tone," *New York Times*, January 22, 2009, A1; Michael D. Shear, "Obama Starts Reversing Bush Policies," *Washington Post*, January 22, 2009, A-1.
2. Gabriel Shoenfeld, "The Bush Secrecy Myth," *Wall Street Journal*, February 22, 2008, A15.
3. Executive Order no. 13,489, 2009 Daily Compilation of Presidential Documents No. 00003, 1–2.
4. Memorandum on the Freedom of Information Act, 2009 Daily Compilation of Presidential Documents No. 00009, 1–2; Memorandum on the Freedom of Information Act 2009, 1–2; Memorandum on Transparency and Open Government, 2009 Daily Compilation of Presidential Documents No. 00010, 1–2.
5. Freedom of Information Act, U.S. Code 5 (1966), §552. Amended (2005).
6. Stolberg, "On First Day," A23.
7. *National Archives and Records Administration v. Favish*, 541 U.S. 157 (2004); *U.S. Department of State v. Ray*, 502 U.S. 164 (1991); *Department of Justice v. Reporters Committee for Freedom of the Press*, 489 U.S. 749 (1989); *Department of State v. Washington Post Co.*, 456 U.S. 595 (1982).
8. *Department of Air Force v. Rose*, 425 U.S. 352 (1976).

9. The FOIA was amended in 1974, 1976, 1986, 1996, and 2007. Pub. L. No. 93-502, 88 Stat. 1561 (1974); Pub. L. No. 94-409, 90 Stat. 1241 (1976); Pub. L. No. 99-570, 100 Stat. 3207 (1986); and Pub. L. No. 104-231, 110 Stat 3048 (1996); Pub. L. No. 110-175, 121 Stat. 2524, §7 (2007).
10. 5 U.S.C. §552(b)(1–9).
11. U.S. Congress. House. 104th Cong., 2nd Sess., 1996. H.R. Rep. No. 104-795, 7; Kent R. Middleton, William Lee, and Bill F. Chamberlin, *The Law of Public Communication* (Boston, MA: Allyn & Bacon, 2011).
12. U.S. Congress. Senate. 89th Cong., 1st Sess., 1965. S. Rep. No. 89-813; U.S. Congress. House. 89th Cong., 2nd Sess., 1966. H.R. Rep. No. 1497.
13. U.S. Congress. Senate. 93rd Cong., 2nd Sess., 1974. S. Rep. No. 93-1200.
14. Association of the Bar of the City of New York, Committee on Federal Legislation, *Amendments to the Freedom of Information Act*, Federal Legislation Report No. 74-1 1974.
15. U.S. Congress. House. 92nd Cong., 2nd Sess., 1972. H.R. Rep. No. 92-1419, 14, 57.
16. H.R. Rep. No. 92-1419 1972, 8.
17. Pub. L. No. 93-502, 88 Stat. 1561 (1974).
18. *FAA v. Robertson*, 422 U.S. 255 (1975).
19. U.S. Congress. House. 94th Cong., 2nd Sess., 1976. H.R. Rep. No. 94-880, 1, 23.
20. Pub. L. No. 104-231 1996.
21. U.S. Congress. Senate. 104th Cong., 2nd Sess., 1996. S. Rep. No. 104-272, 26-27.
22. *DOJ v. Reporters Committee*.
23. H. Rep. No. 89-1497 1966, 2–3, 5–6; *NLRB v. Robbins Tire & Rubber Co.*, 437 U.S. 214 (1978); *DOJ v. Reporters Committee*.
24. Senate Report No. 89-813, 1965.
25. 5 U.S.C. §552(b)(1–9); Senate Report No. 89-813, 1965, 3, 6.
26. Senate Report No. 89-813, 1965, 3.
27. H.R. Report No. 1497, 1966, 6.
28. Alan F. Westin, *Privacy and Freedom* (London: The Bodley Head, 1970).
29. 5 U.S.C. §552 (b)(6).
30. H.R. Report No. 1497, 1966; *Department of State v. Washington Post*.
31. 5 U.S.C. §552 (b)(7)(C)(2005).
32. Freedom of Information Act as Amended in 1974, 120 Congressional Record S9336-37 1974.
33. 5 U.S.C. §552(b)7(C).
34. 5 U.S.C. §552(b)(6).
35. 5 U.S.C. §552(b)(7)(C).
36. *DOJ v. Reporters Committee*.
37. *Air Force v. Rose*.
38. *Air Force v. Rose*, 380–381.
39. *Air Force v. Rose*, 352.
40. *Air Force v. Rose*, 380–381.
41. *Air Force v. Rose*, 380–381, 367–369.
42. *Air Force v. Rose*.
43. *Air Force v. Rose*, 361–362.
44. *Air Force v. Rose*, 361.
45. *FBI v. Abramson*, 456 U.S. 615 (1982), 642.
46. *FBI v. Abramson*, 642.

47. *FBI v. Abramson*, 640–641.
48. *National Archives and Records Admininistration v. Favish*; *Bibles v. Oregon Natural Desert Association*, 519 U.S. 355 (1997); *U.S. Department of Defense v. Federal Labor Relations Authority*, 510 U.S. 487 (1994); *Department of State v. Ray*; *DOJ v. Reporters Committee*; *FBI v. Abramson*; *Department of State v. Washington Post*.
49. *Air Force v. Rose*, 361.
50. *Department of State v. Washington Post*, 600–602.
51. *Department of State v. Washington Post*, 596.
52. *Department of State v. Washington Post*.
53. *Department of State v. Washington Post*.
54. *Air Force v. Rose*, 382.
55. *Department of State v. Washington Post*, 600–602.
56. *Department of State v. Washington Post*.
57. *Department of State v. Washington Post*, 599.
58. *Department of State v. Washington Post*, 599–600.
59. *Department of State v. Washington Post*, 599–600.
60. Lillian R. Bevier, "Information About Individuals in the Hands of Government: Some Reflections on Mechanisms for Privacy Protection," *William & Mary Bill of Rights Journal* 4 (2, 1995): 455.
61. *DOJ v. Reporters Committee*, 749–751.
62. Reporters Committee for Freedom of the Press, founded in 1970, is a nonprofit organization that provides free legal assistance to and on behalf of journalists.
63. *DOJ v. Reporters Committee*, 774; Martin E. Halstuk and Charles N. Davis, "The Public Interest Be Damned: Lower Court Treatment of the Reporters Committee 'Central Purpose' Reformulation," *Administrative Law Review* 54 (Fall 2002): 983.
64. *DOJ v. Reporters Committee*, 772–773.
65. *DOJ v. Reporters Committee*, 774.
66. *DOJ v. Reporters Committee*, 775.
67. *DOJ v. Reporters Committee*, 780–781.
68. *Department of State v. Ray*, 179.
69. *Department of State v. Ray*.
70. *Department of State v. Ray*, 167–168.
71. *Department of State v. Ray*, 168.
72. *Department of State v. Ray*.
73. *Department of State v. Ray*, 170–171, 179.
74. *Department of State v. Ray*, 177–179.
75. *Department of State v. Ray*, 178; citing *DOJ v. Reporters Committee*, 773, quoting *Air Force v. Rose*, 360–361.
76. *Department of State v. Ray*, 179.
77. *Department of State v. Ray*.
78. *Department of State v. Ray*.
79. H.R. Rep. No. 94-880 1976; H.R. Rep. No. 104-795 1996; Senate Rep. No. 104-272 1996; Pub. L. No. 93-502 1974; Pub. L. No. 94-409, 90 Stat. 1241 1976; Pub. L. No. 99-570, 100 Stat. 3207 1986; Pub L. 104-231 1996; Pub. L. No. 110-175 2007; Patrick J. Leahy, "The Electronic FOIA Amendments of 1996: Reformatting the FOIA for On-Line Access," *Administrative Law Review* 50 (Spring 1998): 340; Bevier, "Information About Individuals in the Hands of Government."

80. Michael Hoefges, Martin E. Halstuk, and Bill F. Chamberlin. "Privacy Rights Versus FOIA Disclosure Policy: The 'Uses and Effects' Double Standard in Access to Personally Identifiable Information in Government Records," *William & Mary Bill of Rights Journal* 12 (1, 2003): 1.
81. *National Archives and Records Administration v. Favish*.
82. *National Archives and Records Administration v. Favish*, 173–174.
83. Ruth Marcus, "Clinton Aide Vincent Foster Dies in Apparent Suicide," *Washington Post*, July 21, 1993, A-1; Thomas L. Friedman, "White House Aide Leaves No Clue About Suicide," *New York Times*, July 22, 1993, A-1; *National Archives and Records Administration v. Favish*, 161).
84. Stephen Labaton, "Justice Dept. to Stay on Case of Aide's Death." *New York Times*, July 23, 1993, A-10.
85. Report on the Death of Vincent W. Foster Jr., by the Independent Counsel, In re: Madison Guaranty Savings & Loan Ass'n, to the Special Division of the United States Court of Appeals for the District of Columbia Circuit. Filed July 15, 1997.
86. Brief for Amici Curiae Reporters Committee for Freedom of the Press et al., *Office of Independent Counsel v. Favish*, 217 F.3d 1168 (9th Cir. 2000), 10.
87. *National Archives and Records Administration v. Favish*, 170.
88. *Hale v. U.S. Department of Justice*, 973 F.2d 894 (10th Cir. 1992), 902; *Badhwar v. Department of the Air Force*, 829 F.2d 182 (D.C.Cir. 1987), 185–186; *Marzen v. HHS*, 825 F.2d 1148 (7th Cir. 1987), 1154; *Kiraly v. FBI*, 728 F.2d 273 (6th Cir. 1984), 277–278; *N.Y. Times Co. v. NASA*, 782 F. Supp. 628 (D.D.C. 1991).
89. *National Archives and Records Administration v. Favish*, 172–174.
90. *National Archives and Records Administration v. Favish*, 172.
91. *National Archives and Records Administration v. Favish*, 174.
92. *National Archives and Records Administration v. Favish*, 172.
93. *National Archives and Records Administration v. Favish*, 172.
94. *National Archives and Records Administration v. Favish*, 173–174.
95. *National Archives and Records Administration v. Favish*, 175.
96. Shoenfeld, "The Bush Secrecy Myth," 2008.
97. Executive Order No. 13,489 2009, 1–2; Memorandum on the Freedom of Information Act 2009, 1–2; Memorandum on Transparency and Open Government 2009, 1–2.
98. Memorandum on the Freedom of Information Act 2009, 1–2.
99. Memorandum on Transparency and Open Government 2009, 1–2 at ¶ 2.
100. The Obama Administration's Commitment to Open Government: A Status Report, September 11, 2011.
101. Harry Hammitt, "Obama Memo on FOIA Creates High Expectations," *Access Reports* 35 (2009): 1.
102. Memorandum on the Freedom of Information Act 2009.
103. Executive Order No. 13,489 2009, 1–2.
104. Executive Order No. 13,233, 37 Weekly Compilation of Presidential Documents 1581. 2001.
105. The Presidential Records Act, U. S. Code 44 (1978), §§2201–2207. Amended (2000).
106. Eric Holder, Office of the Attorney General, United States Department of Justice, "Memorandum for Heads of Executive Departments and Agencies," March 19, 2009, http://www.justice.gov/ag/foia-memo-march2009.pdf
107. Holder, "Memorandum for Heads of Executive Departments and Agencies," 2009.
108. Obama Administration's Commitment to Open Government, 2011.

109. Pub. L. No. 104-231, §§1-12, 110 Stat. 3048–3054 (1996). Amending sections of 5 U.S.C. §552; U.S. Congress. Senate. 89th Cong., 1st Sess., 1965. S. Rep. No. 89-813; U.S. Congress. House. 89th Cong., 1st Sess., 1965. H. Rep. No. 89-813, 5.
110. The authors did not include data from the two Executive Branch offices that historically generated the most FOIA requests: the Social Security Administration (SSA) agency and the Department of Health and Human Services (HHS). Their data was not used because SSA and HHS, during the early years of this study, adopted atypical systems for tracking and processing FOIA requests. In fact, their compliance data differed wildly from the rest of the federal government's top departments and agencies. Using their data would have distorted and misrepresented overall FOIA data results. For example, during the six-year period examined, the government received an average of FOIA requests annually totaling in the mid-600,000s range. However, in each of the fiscal years 2006 and 2007, the SSA and HHS reported receiving 18.7 *million* and 19.0 *million* (SSA), and 258,000 and 290,000 requests (HHS), respectively. It was not until the FOIA amendments of 1996 that Congress mandated federal agencies and departments to keep accurate records reporting FOIA compliance. Pub. L. No. 104-231, 110 Stat 3048 (1996). However, it was not until the OPEN Government Act of 2007 before Congress compelled the federal agencies to develop a single uniform system-wide record-keeping procedure (Openness Promotes Effectiveness in our National Government Act of 2007). Pub. L. No. 110-175 (2007). The departments studied: Agriculture, Defense, Homeland Security, Justice, State, Treasury and Labor. The agencies are the Equal Employment Opportunity Commission (EEOC), the National Labor Relations Board (NLRB), the National Archives and Records Administration (NARA), the Securities and Exchange Commission (SEC) and the Environmental Protection Agency (EPA).
111. The official DOJ data for fiscal year 2012 would not be available until 2013.
112. United States Department of Justice. *Freedom of Information Act Guide*. 2010.
113. 5 U.S.C. §552(b)(6 & 7(C)) 2005.
114. Department of Justice, *FOIA Guide*, 2010.
115. *Department of State v. Washington Post.*
116. *DOJ v. Reporters Committee.*
117. *Department of State v. Ray.*
118. *National Archives and Records Administration v. Favish.*
119. *National Archives and Records Administration v. Favish.*

CHAPTER THREE

Public Access and Informational Privacy in Electronic Government Databases

JOEY SENAT

Some forty years ago, a federal district judge conceded even then that computers and electronic databases were "facts of present day life."[1] "Courts can be no more effective than Canute in turning back the tide," wrote Judge Robert L. Carter. "It cannot be contended, at least not seriously, that governmental use of this new technology is constitutionally impermissible." In 1990, the Fourth Circuit Court of Appeals acknowledged that this technology has "provided society with the ability to collect, store, organize, and recall vast amounts of information about individuals."[2] This "information can be useful and even necessary to maintain order and provide communication and convenience in a complex society," the Fourth Circuit said. At the same time, however, the judiciary has recognized that "overzealous data collection and instant data retrieval" pose threats to society.[3]

The central problem for courts, said Chief Justice John W. Fitzgerald of the Michigan Supreme Court, "is to determine how the legal system can best insure [sic] that a proper balance is struck between the traditional libertarian ideals embodied in the concept of privacy and the immense social benefit that computer technology offers."[4]

This book's first two chapters explained some of the U.S. Supreme Court's thinking in balancing informational privacy and public access when government records are compiled in computer databases. This chapter will dig deeper, including how the state courts have viewed the issue, particularly as more and more information is collected by the government about individuals. Whether it's the

NSA or a town in Kansas, when government hoards "big data" on individual citizens, issues increasingly emerge regarding how that information is kept, used, and disseminated.

Overall, courts at the state and federal levels typically treat the release of computerized information in the same way that they treat the release of paper documents—beginning with the recognition that individual privacy and public access are competing interests that are weighed against each other. To determine whether disclosure of personal information contained in government documents would constitute an unreasonable invasion of privacy, courts typically employ some or all of these six factors:

1. The nature and validity of the asserted privacy interest and the degree of the invasion of that interest;
2. The extent or value of the public's interest in disclosure;
3. The purpose or objective of the requester seeking disclosure;
4. The availability of the information from other sources;
5. Whether the government promised confidentiality; and
6. Whether it is possible to redact personal information so as to limit the breach of individual privacy.

This chapter will walk through these six factors, explaining how the courts have negotiated through the sticky bog to balance individual privacy with the public's ability to get information about its government.

THE PRIVACY INTEREST AT STAKE

Courts considering public access to government databases begin their balancing by determining the privacy interest at stake for individuals contained in those data. Most use a two-pronged approach in which they decide if disclosure would constitute an invasion of privacy and, if so, the degree or seriousness of that invasion. Key questions include: What is privacy? Does the individual to whom the privacy relates have a reasonable expectation or legitimate expectation of confidentiality? Would release of the information cause harm to the individual? Would a reasonable person consider release of the information "highly offensive"?

Few courts attempt to define privacy. Michigan Chief Justice John W. Fitzgerald noted: "The concept of privacy is elusive. Social scientists and legal scholars alike have struggled for a definition expansive enough to include important concerns and yet narrow enough to be workable."[5] He concluded, however, that "[a]s society has expanded and distance contracted because of advances in communication and travel, the right to privacy for many has become the ability to choose with whom and under what circumstances they will communicate."[6]

Likewise, California Chief Justice Rose Elizabeth Bird noted that a 1972 privacy amendment to her state's constitution "protects the right to informational privacy."[7] Quoting from an election brochure argument supporting the amendment, Justice Bird wrote: "Fundamental to our privacy is the ability to control circulation of personal information....This is essential to social relationships and personal freedom. The proliferation of government and business records over which we have no control limits our ability to control our personal lives."[8]

Writing for the U.S. Supreme Court in 1989 for the *Reporters Committee* case, Justice John Paul Stevens relied upon a definition from Webster's Third New International Dictionary[9] and definitions provided by two privacy advocates—Adam Breckenridge[10] and Alan Westin[11]—for support that privacy encompasses the individual's right to control the flow of personal information.[12] Stevens rejected the argument that the privacy interest approached "zero" just because the information was available to the public elsewhere, calling that a "cramped notion of personal privacy."[13] In the same case, however, the U.S. Court of Appeals for the District of Columbia had used the same dictionary definition to come to a different conclusion about the privacy interest at stake in an FBI computer rap sheet compiled from public records, holding that the "ordinary meaning of privacy suggests that (FOIA) Exemption 7(C) does not exempt records consisting of information that is publicly available."[14]

It was the U.S. Supreme Court's description of privacy that was relied upon by the Arizona Supreme Court in 1998. In holding that broadcast journalists could not access a school district's computer records containing teachers' dates of birth, the court said, "Although we have never defined the meaning of privacy under the Public Records Law, the (U.S.) Supreme Court, interpreting the FOIA, has stated that information is 'private if it is intended for or restricted to the use of a particular person or group or class of persons: not freely available to the public.'"[15] The Arizona court added that the U.S. Supreme Court in *Reporters Committee* had "stated that the privacy interest encompasses 'the individual's control of information concerning his or her person.'"[16]

Rather than defining the concept of privacy, most courts deciding cases involving access to computer records tend to evaluate the privacy interest at stake based on the information's content and/or context. However, attitudes toward computers have played a key role in some cases. Judges in some cases viewed the influence of computerized data on individual privacy very differently even when faced with the same facts.

For example, the Michigan Supreme Court split evenly over whether the computer format for data affected the individual's reasonable expectation of privacy.[17] At issue was a copy of the magnetic computer tape that Michigan State University used to produce its student directory. The requester wanted to develop a list from which he could make mailings on behalf of political parties who would pay him for that service. Michigan State officials refused to provide the tape, offering instead a copy of the student directory when it was published or an immediate printout of the information on the magnetic tape.

A divided Michigan Supreme Court affirmed the lower court ruling that disclosure of the computer tape would constitute an invasion of privacy. Chief Justice Fitzgerald, joined by two other justices, concluded that the release of names and addresses on a magnetic tape was a more serious invasion of privacy than disclosure of the information in a paper directory. He acknowledged that students could have opted out of being included in the student directory. He also acknowledged that students who did not opt out "should have known" that the information was available to the public and could be changed to a computer form by anyone in the public to compile mailing lists. "However, it does not follow that students should have known that an efficient and intrusive computer mailing system already was available to anyone for a nominal sum," he contended. "In deciding whether to appear in or opt out of the directory, students should not have been expected to consider the mechanics by which the university published the information."[18]

In contrast, Justice James Ryan, joined by two other justices, disagreed that students had any reasonable expectation that their information would only be released in printed form. He contended that no reasonable expectation of privacy exists if the information sought is a computer copy of a public record also available in paper. In other words, if access to the paper record is not a violation of privacy, then access to a computer copy of the same record is not a violation. "We cannot accept the conclusion that the Legislature intended to allow a public body to exempt otherwise public records from disclosure by the simple expedient of converting the public record from one form to another," Justice Ryan reasoned. "Surely such a result would exalt form over substance. The plain language of the statute reveals a legislative intent to treat all government 'writings' in the same fashion regardless of form."[19]

Courts in New Jersey,[20] New Mexico,[21] and New York[22] have used the same reasoning as Justice Ryan to find no privacy interest exists in the computer format of records if they are public as paper documents. The New Jersey Supreme Court, however, emphasized that its decision could not be generalized to all cases in which computer copies of public records were sought. Computer tapes could be "far more revealing" as well as create "the potential for far more intrusive inspections," it explained. "Unlike paper records, computerized records can be rapidly retrieved, searched, and reassembled in novel and unique ways, not previously imagined. For example, doctors can search for medical-malpractice claims to avoid treating litigious patients; employers can search for workers'-compensation claims to avoid hiring those who have previously filed such claims; and credit companies can search for outstanding judgments and other financial data. Thus, the form in which information is disseminated can be a factor in the use of and access to records," the court said.[23]

Only Justice Ryan's opinion in Kestenbaum explicitly rejected the notion that computers automatically pose a significant threat to privacy in the context of names and home addresses. "An individual's home may still be a 'castle' into which 'not even the king may enter,' but nothing prevents the king or anyone else from telling others whose castle it is, particularly when the castle-dweller himself has voluntarily released that information to the general public," Justice Ryan reasoned.[24] He said the "supposed protection of students' privacy gained by denying" access to the computer tape was unfair because it penalized only those groups unable to afford converting the printed copy into an electronic format and was illusory because a commercial entity would be economically motivated to sell the computer version to as many groups as possible to recoup the cost of conversion.[25]

In contrast, other judges expressed the belief that computerized information poses a greater threat to individual privacy than paper copies do. For example, Chief Justice Fitzgerald said, "Form, not just content, affects the nature of information. Seemingly benign data in an intrusive form takes on quite different characteristics than if it were merely printed. The very existence of information in computer-ready format may serve to motivate an invasion of privacy."[26]

Some courts are fearful of private databases of personal information compiled in part from public records. For example, the Massachusetts Court of Appeals said privacy is threatened even when non-intimate details about large numbers of people are placed in computer databases. "There is a negative public interest in placing the private affairs of so many individuals in computer banks available for public scrutiny," said the court.[27] It noted that a 1937 Massachusetts Supreme Court decision declaring motor vehicle records open "was written in an era prior to the advent of modern data processing technology which permits 'the aggregation of pieces of personal information into large central data banks.'"[28]

The Oklahoma Supreme Court feared that a database of birthdates of state employees could be used for identity theft and scams as well as to find other information. "With both a name and a birth date, one can obtain information about: an individual's criminal record; arrest record (which may not include disposition of the charges); driving record; state of origin; political party affiliation; social security number; current and past addresses; civil litigation record; liens; property owned; credit history; financial accounts; and, quite possibly, information concerning an individual's complete medical and military histories; and insurance and investment portfolio," the court said.[29]

The possibility that information about children could be published worldwide on the Internet was a key consideration of the Ohio Supreme Court when it denied access to an electronic database that included the names, home addresses, family information, emergency contact information, and medical history information of children who had received photographic identification cards to use a city's pools and recreation facilities. The court said the database was not a public record because it represented personal information collected by government that did not shed light on government activities.

But the Ohio court also held that even if the database were a public record, its disclosure would constitute an unwarranted invasion of privacy in The Information Age because "any perceived threat...no matter how attenuated, cannot be discounted." The court explained: "Technological advances have made...it possible to generate and collect vast amounts of personal, identifying information through everyday transactions such as credit card purchases and cellular telephone use. The advent of the Internet and its proliferation of users has dramatically increased, almost beyond comprehension, our ability to collect, analyze, exchange, and transmit data, including personal information. In that regard, it is not beyond the realm of possibility that the information at issue herein might be posted on the Internet and transmitted to millions of people." Because of the inherent vulnerability of children, the court said it was "necessary to take precautions to prevent, or at least limit, any opportunities for victimization." Therefore, the court could not "in good conscience" release the information.[30]

Other courts declared that computers pose a threat to privacy not only because they provide current information but also because they overcome practical obscurity by helping create life-long dossiers pieced together from data previously scattered among far-flung sources. The U.S. Supreme Court, for example, said in *Reporters Committee* that the privacy interest in a computerized criminal rap sheet compiled by the FBI was "affected by the fact that in today's society the computer can accumulate and store information that would otherwise have surely been forgotten long before a person attains age 80, when the FBI's rap sheets are discarded....Plainly there is a vast difference between the public records that might be found after a diligent search of courthouse files, county archives, and local police stations throughout the country and a computerized summary located in a single clearinghouse of information"[31]

In contrast, the D.C. Circuit had rejected the argument that access to a computerized rap sheet should be denied because computers made the records available too long. The court wrote: "We see no principled basis by which a court can determine that a crime is so 'minor' that information regarding it, which a state considered significant enough to place on the public record, is in reality of little public interest. Nor can we say that an older public record has lost its public interest—old records may have historical importance."[32]

But some state courts have adopted the Supreme Court's reasoning. The Louisiana Court of Appeals, for example, relied upon the *Reporters Committee* opinion when it declared the state's centralized, computer-based criminal justice information system off limits to the public.[33] The California Court of Appeals likewise relied upon it when denying access to computer tapes of a county court system's compilation of criminal offense information that included the name, birthdate, and zip code of every person against whom criminal charges were pending, plus the case number, date of offense, charges filed, pending court dates, and disposition.[34]

Robert Westbrook had told the California trial court that not having the computer tapes would require him to travel to 46 municipal courts and "(a)s a result, no one would be able to afford what he would have to charge them for the information." The trial court declared as "nonsensical" the government's argument that Westbrook could have some information on a computer tape but obtain the rest only by traveling to each court. The appellate court, however, overturned that decision, reasoning that there exists "a qualitative difference between obtaining information from a specific docket or on a specified individual, and obtaining docket information on every person against whom criminal charges are pending in the municipal court." Noting that the aggregate nature of the data made it valuable to Westbrook, the appellate court said, "(I)t is that same quality which makes its dissemination constitutionally dangerous."[35] Fearing the power of computers to eliminate practical obscurity, the court said Westbrook had "information from which he can, over the years, compile his own private data base of criminal offender record information."[36] "(T)he potential for misuse of the information is obvious," the court said. "If, for example, the court ordered a record maintained by a criminal justice agency to be sealed or destroyed because a defendant had been found to be factually innocent of the charges, the information would still be available for sale by (Westbrook). The only control on access to the information in (Westbrook's) possession would be the price he places on it."[37]

The Colorado Supreme Court relied upon the reasoning in *Westbrook* and *Reporters Committee* when it agreed that computerized compilations of court records should be treated differently from individual case files and that such requests should be decided on a case-by-case basis. "Whether bulk data should be released and to whom is a matter of important policy that necessarily involves the balancing of individual privacy concerns, public safety, and the public interest in fair and just operation of the court system," said the court.[38]

These cases indicate that at least some judges are more willing to accept—without citing any social scientific support or evidence beyond their own speculation—that computers and computerized information threaten individual privacy more than paper records do. The next factor to explore is whether courts believe computers contribute to the public interest against which the privacy interest is balanced.

THE PUBLIC INTERESTS SERVED BY DISCLOSURE

Courts dealing with computer records typically weigh against the individual's privacy interest the good that disclosure would bring to the general public. In other words, the determination of whether an invasion of privacy is unwarranted can depend upon the public interest at stake in the request.

In *Reporters Committee*, however, as previously discussed in this book, the U.S. Supreme Court narrowed the public interest under the Freedom of Information Act to only disclosures of "(o)fficial information that sheds light on an agency's performance of its statutory duties."[39] Personal information in the hands of government that did not shed light on the conduct of government would not meet that public-interest standard. While a rap sheet conceivably could provide details for a news story, Justice Stevens said: "(I)n itself, this is not the kind of public interest for which Congress enacted the FOIA. In other words, although there is undoubtedly some public interest in anyone's criminal history, especially if the history is in some way related to the subject's dealing with a public official or agency, the FOIA's central purpose is to ensure that the Government's activities be opened to the sharp eye of public scrutiny, not that information about private citizens that happens to be in the warehouse of the Government be so disclosed."[40]

State courts have adopted the Supreme Court's reasoning when determining the public interest under their respective public records statutes. For example, the Louisiana Court of Appeals, noting that its statute was similar to the federal FOIA, held that the policy behind its statute was "the public's right to be informed of what our government is up to."[41] The Oklahoma Supreme Court likewise used the narrow description of public interest when it denied public access to state employees' birthdates under that state's Open Records Act. "There simply is no instance in which we can fathom how such information would advance the public's interest in assuring that the government is properly performing its function," the court explained.[42]

A few judges have noted that computers can aid in the public's inspection of government records and, therefore, in the disclosure of what government is doing. For example, the New Hampshire Supreme Court reasoned that releasing computer copies of real estate tax assessment records made more sense than restricting disclosure to paper copies. After noting that examining 35,000 real estate tax assessment records would take 200 days at a cost of about $10,000, the court said, "The ease and minimal cost of the tape reproduction as compared to the expense and labor involved in abstracting the information from the field cards are a common sense argument in favor of the former."[43]

The New Mexico Supreme Court similarly held that the right to inspect public records should "carry with it the benefits arising from improved methods and techniques of recording and utilizing the information...so long as proper safeguards are exercised as to their use, inspection, and safety."[44] And in Kestenbaum, Justice Ryan of the Michigan Supreme Court argued that public access to electronic records should not be denied just because computer copies make the information more usable, saying that "to equate usefulness with intrusiveness is to turn the FOIA on its head."[45]

But these judges were clearly in the minority when they recognized that computerized information could help the public learn about the actions of government. Most courts seem to place little value on the use of computers to facilitate access to public records and, therefore, to support a public interest in disclosure.

However, the computer format of the public records is very much on the minds of some judges as they consider the private interests actually—or potentially—served by the release of the documents.

THE PRIVATE INTERESTS SERVED BY DISCLOSURE

A number of federal and state courts considering claims to government computer files reaffirmed the principle that any person is eligible to request public records. In *Reporters Committee*, the U.S. Supreme Court held that determining whether an invasion of privacy is warranted "cannot turn on the purposes for which the request for information is made" and that "the identity of the requesting party has no bearing on the merits of his or her FOIA request."[46]

State courts in California,[47] Illinois,[48] Kansas,[49] Kentucky,[50] Louisiana,[51] and Massachusetts[52] have used the same reasoning when ruling on access to computer copies of public records. An Illinois court, for example, noted that its state public records statute did "not require that the persons requesting the information explain their need for that information or their planned use of the information. The Act seeks to achieve a highly desirable goal; namely, that the public knows how its tax dollars are being spent."[53] A California court reasoned that the requester's purpose could not be considered because "once a public record is disclosed to the requesting party, it must be made available for inspection by the public in general."[54]

However, a number of courts have considered the private interest served by disclosure, and some have been hostile to the commercial motivations of requesters.[55] In many of these cases, the records probably would not have been requested had they been paper files. For example, the California Court of Appeals noted that the entrepreneurial reason for seeking bulk court records would not have existed had the information not been in computer files.[56] Under the penal code, the court said, a business selling criminal background information to the public was only entitled to criminal offense information compiled by a court if the company could show a "compelling need."[57] In denying access, the court said selling the information did not qualify as a legally acceptable need to know the information.[58]

Other courts were equally hostile to hypothetical private interests that could be better served by computer records. For example, the New Jersey Supreme Court explained that computer records made it easier for doctors to search medical-malpractice claims for patients more likely to sue and for employers to identify job applicants who had filed workers'-compensation claims.[59] The Ohio Supreme

Court worried that a government computer file of personal data about children could be placed on the Internet and be used for criminal purposes.[60] Oklahoma Supreme Court justices worried that a database of state employees' birthdates could be used to help criminals commit identity theft and scams.[61]

While the computer format of records was not the explicit reason courts devalued the private interests of some requesters, it certainly was a consideration by those courts.

The computer format is a key consideration in courts' analyses of only one other of the six factors often used in balancing access and privacy claims. The next section of this chapter examines the role that computerized information plays when courts consider whether to redact exempt information and allow disclosure of non-exempt data.

THE POSSIBILITY OF REDACTING PERSONAL INFORMATION

Some courts are supportive of redaction of personal information when computer records are at issue. For example, a Connecticut trial court said the redaction of exempted names and addresses from computerized Department of Motor Vehicle records would protect the privacy of individual motorists and still allow the disclosure of important information. "Indeed," the judge said, "such sanitized information could be the source of important statistics, useful for a variety of legitimate purposes, without harming any of the interests protected by the statute."[62]

Redaction is sometimes required by state statute. Illinois courts, for example, noted their state public records statute requires any agency maintaining a record with exempt and nonexempt data to separate the exempt material and disclose the rest of the document.[63] Therefore, the Illinois Supreme Court said, "The mere presence or commingling of exempt material does not prevent the district from releasing the nonexempt portion of the record."[64]

Several courts recognized that, compared to other record formats, computerized data provide more protection for confidential information. For example, the Minnesota Supreme Court said, "Retrieving the data from the computer rather than allowing access to the microfilm copies protects the confidentiality of the patients since only the specific information sought need be disclosed."[65] Kansas[66] and Illinois[67] courts were willing to require governmental entities to create new computer software if necessary to delete confidential information. The courts required the requesters to pay for the special programs.

Illinois[68] and New York[69] courts even ordered school districts to "scramble" computer databases of standardized student test scores to further cloak identifying information. But Illinois Supreme Court Justice Benjamin Miller disagreed with this approach, saying he saw nothing in the public records statute "which indicates that the legislature intended to impose a duty on public bodies to use their computer capabilities to provide information in a form that would make the material nonexempt. The act simply does not differentiate between records stored in computers and those maintained manually."[70] Justice Miller also said he was not "convinced that such a distinction would be advisable" because it would create a two-tier system for disclosure that encouraged government agencies to keep documents in a paper format with fewer disclosure requirements.[71]

But overall, courts seem more likely to accept redaction as a way to protect confidential information while disclosing non-exempt data. Computers are likely to be viewed as a useful tool in achieving that goal. In essence, a computerized format for government records makes it easier for courts to withhold private information while disclosing the remaining public information.

THE REMAINING FACTORS

The computer format does not matter to courts when they consider the two remaining factors. Courts weigh these factors in the same ways that courts do when the records are made of paper.

Courts sometimes consider whether alternate means of obtaining the information reduces the need for the requested records to be disclosed. Courts in Ohio and Oklahoma, for example, said the availability of the information from other sources did not "dissolve the individual's interest in controlling the dissemination of information regarding personal matters."[72]

"(T)he mere fact that the state distributes records containing home addresses to state-employee unions and certain nongovernmental vendors or that some addresses are available through other public records, e.g., voter registration and county property records, does not extinguish state employees' privacy interests in those addresses," the Ohio Supreme Court said.[73]

But this factor is considered without discussion of computer copies versus hard copies of the information. The key is simply whether the information itself is available elsewhere. For example, the California Court of Appeals decided that *Mercury News* reporters seeking the identities of people who had complained to the city about municipal airport noise had "alternative means of contacting and interviewing the complainants other than by intruding on their privacy through forced disclosure of their identities from government records." Reporters could canvass neighborhoods near the airport and could contact complainants who had made their identities public by appearing at city council meetings, joining an anti-airport noise group, or by disclosing their names on the group's World Wide Web site, the court said.[74]

Computers also are a non-issue in the consideration of whether an agency's promise of confidentiality overrides a statutory duty to disclose public records.

The Arizona Court of Appeals, for example, said a school district's promise of confidentiality to teachers could not preclude the release from computer databases of the names and birth dates for some 30,000 teachers. "(I)f the promise of confidentiality were to end our inquiry, we would be allowing a school district official to eliminate the public's right under Arizona's Public Records Law," said the court. "We cannot allow a school district to exempt public records from disclosure simply by promising confidentiality."[75]

CONCLUSION

Courts typically treat the release of computer copies of public records in the same way that they treat the release of paper copies. Most courts make only passing references to the computer format of the information. Courts apply no new factors when weighing the competing claims of privacy and public access to computer records, relying instead on some or all of six factors used when paper documents are requested. However, judicial attitudes toward computers can sometimes serve a key role when courts determine the privacy interest at stake. For example, some judges assume that people have a reasonable expectation that public records will not be made available in a computer-ready format. Other judges contend that no such expectation exists if the computer data being sought is also available in paper.

Some judges believe that computer files simply pose more of a threat to privacy than do paper files. Simply put, they seem to fear the technology. In their eyes, computer databases of even benign personal information make the information too long lasting, too usable by third parties and, therefore, too intrusive to the individual. These fears and assumptions are given the weight of law without the support of stronger evidence, such as a statistical analysis regarding the actual possibility of privacy-invading actions by those who might obtain the information. Without a statutory definition of informational privacy and a firmer set of guidelines to follow, judges are relying upon an "I-know-it-when-I-see-it" and even an "I-know-it-when-I-fear-it" approach to determining the privacy interest at stake.

But while some courts have been quick to accept that disclosure could ultimately lead to a violation of privacy, courts typically reject a similar logic that disclosure, in addition to meeting the requester's needs, also could eventually shed light on government operations. Rarely do judges see computers as a tool to help the public learn what government is doing, which courts have generally considered to be the most important public interest to be weighed against individual privacy.

The computer format of records also seems to work against requesters with strictly private, particularly commercial, interests in government records. Businesses face two obstacles when requesting copies of computer files: judicial resistance to allowing public records to foster commercial gain and judicial suspicions toward computers being used to compile and manipulate databases of personal information.

For the purposes of evaluating potential privacy violations, courts take seriously the computer's capability to accumulate, store, manage and distribute great amounts of data. Yet few judges note that those same capabilities can help the public inspect government records and, therefore, ultimately understand what the government is up to. Courts seem to be ignoring valuable traits of the computer that support this public interest in disclosure.

At the same time, however, courts do not seem uninformed about what computers can do. Judges are willing to use the computer to redact personal information from government data and allow a limited disclosure. Courts have even ordered government to "scramble" alphabetized computer data to further disguise identifying information. Redaction, though, can also be a compromise solution that allows courts to avoid hard questions about the private nature of information and the right of the public to see and use that data.

Overall, courts ignore the fact that the requested public records are in a computer-ready format. For the most part, judges treat computer copies of government documents no differently than they do paper copies. They apply the same factors in the same ways. However, when computers are explicitly considered as part of the formula, the balancing of these factors is clearly weighted in favor of privacy.

For privacy advocates, this means they are more likely to find an ally on the bench when the battle over privacy and public access enters the courtroom. For journalists and others in the public seeking access to government records, this means their battles should be waged in Congress and state legislatures, where they can argue for statutes articulating the same treatment for computer records as for paper ones.

NOTES

1. *Roe v. Ingraham*, 357 F. Supp. 1217 (S.D.N.Y. 1973), 1222.
2. *Walls v. City of Petersburg*, 895 F.2d 188 (4th Cir. 1990), 194–95.
3. *Roe v. Ingraham*, 1222; *Walls v. City of Petersburg*, 194–95.
4. *Kestenbaum v. Michigan State Univ.*, 327 N.W.2d 783 (Mich. 1982), 789.
5. *Kestenbaum v. Michigan State Univ.*, 785.
6. *Kestenbaum v. Michigan State Univ.*, 786.
7. *Perkey v. DMV*, 721 P.2d 50 (Calif. 1986).
8. *Perkey v. DMV*, 58.
9. *Webster's Third New International Dictionary* (Springfield, MA: Merriam Webster 1976).
10. Adam C. Breckenridge, *The Right to Privacy* (Lincoln, NE: University of Nebraska Press, 1970), 1.
11. Alan F. Westin, *Privacy and Freedom* (London, England: The Bodley Head, 1967), 7.

12. *Reporters Comm. for Freedom of the Press v. Department of Justice*, 816 F.2d 730 (D.C. Cir. 1987), rev'd sub nom. *Department of Justice v. Reporters Comm. for Freedom of the Press*, 489 U.S. 749, 109 S. Ct. 1468, 103 L.Ed. 2D 774 (1989), 763–64.
13. *Reporters Committee v. DOJ*, 762–63.
14. *Reporters Committee v. DOJ*, 738.
15. *Scottsdale Unified School District No. 48 of Maricopa County v. KPNX Broadcasting Co.*, 937 P.2d 689, 691–92 (Ariz. Ct. App. 1997), *vacated*, 955 P.2d 534 (Ariz. 1998) (vacated on other issues), 538.
16. *Scottsdale Unified School District v. KPNX*, 538.
17. *Kestenbaum v. Michigan State Univ.*
18. *Kestenbaum v. Michigan State Univ.*, 789.
19. *Kestenbaum v. Michigan State Univ.*, 802.
20. *Higg-A-Rella, Inc. v. County of Essex*, 647 A.2d 862, 865 (N.J. Super. Ct. App. Div. 1994), *affiliated*, 660 A.2d 1163, 1170 (N.J. 1995).
21. *Ortiz v. Jaramillo*, 483 P.2d 500 (N.M. 1971).
22. *Szikszay v. Buelow*, 436 N.Y.S.2d 558 (N.Y. Sup. Ct. 1981).
23. *Higg-A-Rella v. County of Essex*, 1170–71.
24. *Kestenbaum v. Michigan State Univ.*, 796.
25. *Kestenbaum v. Michigan State Univ.*, 801.
26. *Kestenbaum v. Michigan State Univ.*, 789.
27. *Doe v. Registrar of Motor Vehicles*, 528 N.E.2d 880 (Mass. App. Ct. 1988), 886.
28. *Doe v. Registrar of Motor Vehicles*, 884.
29. *Oklahoma Public Employees Association v. State ex rel. Oklahoma Office of Personnel Management*, 2011 OK 68, 32, 2011 Okla. LEXIS 63, 39 Media L. Rep. (BNA) 2610 (Okla. 2011).
30. *State ex rel. McCleary v. Roberts*, 725 N.E.2d 1144 (Ohio 2000), 1149.
31. *Reporters Committee v. DOJ*, 771.
32. *Reporters Committee v. DOJ*, 741.
33. *Ellerbe v. Andrews*, 623 So. 2d 41 (La. Ct. App. 1993).
34. *Westbrook v. County of Los Angeles*, 32 Cal. Rptr. 2d 382 (Cal. Ct. App. 1994), *review denied*, 1994 Cal. LEXIS 5772 (Cal. 1994).
35. *Westbrook v. County of Los Angeles*, 387.
36. *Westbrook v. County of Los Angeles*, 384.
37. *Westbrook v. County of Los Angeles*, 387.
38. *Office of the State Court Administrator v. Background Information Services*, 994 P.2d 420 (Colo. 1999), 429–30.
39. *Reporters Committee v. DOJ*, 772.
40. *Reporters Committee v. DOJ*, 774.
41. *Ellerbe v. Andrews*, 44–45.
42. *Oklahoma Public Employees Association v. State ex rel. Oklahoma Office of Personnel Management*, 35.
43. *Menge v. City of Manchester*, 311 A.2d 116 (N.H. 1973), 119.
44. *Ortiz v. Jaramillo*, 501.
45. *Kestenbaum v. Michigan State Univ.*, 802.
46. *Reporters Committee v. DOJ*, 771.
47. *City of San Jose v. Superior Court of Santa Clara County*, 74 Cal. App. 4th 1008 (Cal. Ct. App. 1999).
48. *Family Life League v. Department of Public Aid*, 478 N.E.2d 432 (Ill. App. Ct. 1985).
49. *State ex rel. Stephan v. Harder*, 641 P.2d 366, 8 Media L. Rep. (BNA) 1891 (Kan. 1982).
50. *Zink v. Commonwealth*, 902 S.W.2d 825 (Ky. Ct. App. 1994).

51. *Webb v. City of Shreveport*, 371 So. 2d 316, 5 Media L. Rep. (BNA) 1729 (La. Ct. App. 1979), *writ denied*, 374 So. 2d 657 (La. 1979).
52. *Doe v. Registrar of Motor Vehicles*.
53. *Family Life League v. Department of Pub. Aid*, 1057–58.
54. *City of San Jose v. Superior Court of Santa Clara County*, 1018.
55. See, for example, *Pantos v. San Francisco*, 198 Cal. Rptr. 489, 10 Media L. Rep. (BNA) 1279 (Cal. Ct. App. 1984).
56. *Westbrook v. County of Los Angeles*, 387.
57. *Westbrook v. County of Los Angeles*, 384–85.
58. *Westbrook v. County of Los Angeles*, 386.
59. *Higg-A-Rella v. County of Essex*.
60. *State ex rel. McCleary v. Roberts*.
61. *Oklahoma Public Employees Association v. State ex rel. Oklahoma Office of Personnel Management*.
62. *Kozlowski v. Freedom of Info. commission*, 1997 Conn. Super. LEXIS 2000 (Conn. Super. Ct. 1997); see, also, *Minnesota Medical Association v. Minnesota Department of Public Welfare*, 274 N.W.2d 84, 4 Media L. Rep. 1872 (Minn. 1978).
63. *Bowie v. Evanston Community Consolidated School District. No. 65*, 538 N.E.2d 557 (Ill. 1989); *Hamer v. Lentz*, 525 N.E.2d 1045 (Ill. Ct. App. 1988), *affiliated*, 547 N.E.2d 191, 17 Media L. Rep. (BNA) 1268 (Ill. 1989).
64. *Bowie v. Evanston*, 560.
65. *Minnesota Medical Association v. Minnesota Department of Public Welfare*, 86.
66. *Family Life League v. Department of Public Aid*, 1058.
67. *Hamer v. Lentz*, 195.
68. *Bowie v. Evanston*.
69. *Kryston v. Board of Education*, 77 A.D.2d 896 (N.Y. App. Div. 1980).
70. *Bowie v. Evanston*, 563.
71. *Bowie v. Evanston*, 563–64.
72. *Oklahoma Public Employees Association v. State ex rel. Oklahoma Office of Personnel Management*, 34.
73. *State ex rel. Dispatch Printing Co. v. Johnson*, 833 N.E.2d 274 (Ohio 2005).
74. *City of San Jose v. Superior Court of Santa Clara County*, 1025.
75. *Scottsdale Unified School District v. KPNX*, 691–92.

CHAPTER FOUR

Conflict in a Digital World: The European Context

CHERYL ANN BISHOP

To foster greater transparency, in 2009 the Spanish government placed online virtual reams of official records dating back at least 350 years, including criminal pardons, bankruptcy records, and civil service exam results. All of them could be found through a simple Google search.[1] Following publication, Spain's data protection agency received nearly 100 complaints from individuals identified in the public documents and subsequent news reports. Under the Spanish Data Protection Authority, citizens can sue to have sensitive information pertaining to them removed from online. One of the complaints came from a leading surgeon who was acquitted of criminal negligence more than a decade earlier, yet a Google search of his name produced the original arrest record but not the acquittal.[2]

This type of legal recourse is often referred to as the "right to be forgotten" and is understood as an information privacy right, also known as "data protection." According to David Banisar, information privacy "encompasses the right of individuals to control personal information such as financial and medical information held by other parties and the creation of rules governing the collection and handling of this information."[3] The "right to be forgotten" expands this right to allow individuals in some contexts to demand removal of sensitive personal information.

In a novel move, Spain's data protection agency ordered Google to remove links to the offending information in the public records and in online news articles. Google, which is appealing the orders, warned that this case could have a "profound, chilling effect" on freedom of expression. The case is in front of the European Court of Justice (ECJ) for clarification regarding European Union (EU) law on data protection.[4]

This case highlights the escalating tension between the right to information privacy and rights to free expression and information in a digitized world. Balancing these competing interests is not just a focus of the U.S. Supreme Court, and judges throughout the world are deriving concepts and solutions from which we can all learn. New technologies provide instantaneous communication and access to information. The global trend to conceptualize access to government information as a public good, and even a human right,[5] has sparked government agencies around the world to provide access to records and government databases online. More and more nongovernmental organizations are taking advantage of the right to government information and providing access to government documents online. Those with an Internet connection also have access to the growing number of online archives providing everything from cute cat photos to national archives across the globe. And thanks to social media networking sites, individuals can and do provide unparalleled levels of personal information to online communities and beyond.

These new technologies also enable governments and the private sector to amass and analyze unprecedented quantities of data, much of which contains our personal information.[6] Every five years, the amount of digital information increases tenfold. According to the technology company Cisco, by the end of 2016, "the gigabyte equivalent of all movies ever made will cross the global Internet every three minutes."[7] Tene and Polonetsky, writing in the Stanford Law Review Online, argue that "data has become the raw material of production, a new source of immense economic and social value."[8] Aggregation and processing of these data can have immense benefits to society, but also create public concern about privacy and raise crucial legal questions regarding who can access and control these data.

These legitimate concerns about information privacy have resulted in more than 60 countries passing data protection laws.[9] The EU has been at the forefront of addressing data protection issues. In 1995, it passed one of the more stringent data protection laws in the world, the EU Data Protection Directive, which applies not only to personal information held by the private sector, but also information held by governments; in other words, public records.[10]

Currently, the EU is in the process of revisiting its data protection framework, which privacy experts have referred to as "the world's broadest and most stringent,"[11] to provide more expansive data protection rights, especially in the online context. In January 2012, the European Commission issued a draft Data Protection Regulation to replace the Directive.[12] The proposed Regulation, which includes the controversial "right to be forgotten," has sparked intense debate regarding balancing the right to information privacy with rights of freedom of expression and access to information.

Adam Thierer, writing for Forbes magazine, voiced his concern bluntly: "What we are talking about here is the destruction of history, otherwise known as censorship. Few would have suggested that burning books was a smart way to protect privacy in the past. Is burning binary bits of information any wiser?" He goes on to question whether public figures could claim "a right to be forgotten" when "a journalist pens an article about them beating their wife or committing corporate fraud."[13] Google's Global Privacy Counsel Peter Fleischer warned that such a right could make Internet search engines "censors-in-chief of the world's web content," obligating them "to delete the content, regardless of whether it was true or fair or legal, regardless of who published it, and regardless of the fact that the search companies had nothing to do with the content."[14]

In light of these concerns, the purpose of this chapter is to provide an analysis of how rights to free expression and information are balanced with the right to information privacy set out in the current Data Protection Directive and the proposed Regulation, and attempts to bring clarity to the "right to be forgotten" debate. This chapter begins by providing a brief discussion of the European framework regarding the right to information privacy. It then offers an overview of the current Directive and proposed Regulation. Next, it discusses the current Directive regarding implications for freedom of expression and government transparency. The final section examines these implications in light of the proposed "right to be forgotten" in the draft Regulation.

THE EUROPEAN CONCEPTUALIZATION OF PRIVACY

In the European context, the right to privacy is understood as a fundamental human right. Article 8 of the European Convention on Human Rights states in part, "Everyone has the right to respect for his private and family life, his home, and his correspondence."[15] According to privacy scholar James Whitman, the European approach to privacy is about "a right to respect and personal dignity," which include "rights to one's image, name, and reputation" the "right to informational self-determination—the right to control the sorts of information disclosed about oneself." Whitman argues that these rights are related to the control of one's public image and how one is perceived by others. They help protect against unwanted public exposure, embarrassment, or humiliation. This understanding of privacy is very suspicious of entities that gather and disseminate information due to the potential harms to one's public dignity.[16]

In regards to information privacy, Europeans have been particularly sensitive regarding the collection of personal data by governments due to the history of authoritarian control such as the Nazis' Gestapo and the Soviets' KGB.[17] The right to information privacy also is understood as a fundamental human right. The EU Charter of Fundamental Rights enshrines data protection as a fundamental right separate from the right to private and family life. According to the Charter, "Everyone has the right to the protection of personal data concerning him or her."[18]

According to a report of the EU Agency of Fundamental Rights, the inclusion of data protection in the Charter "is a recognition by the EU of the importance of technological progress, and an attempt to make sure that fundamental rights take account of this progress. The undeniable fact that our lives are now becoming a continuous exchange of information, and that we live in a continuous stream of data, means that data protection is gaining importance and moving to the centre of the political and institutional system."[19]

The European Court of Human Rights has a long line of case law regarding information privacy, particularly the right of individuals to obtain information held about them from their governments. For example, in 1987, the Court held that the storing of information relating to an individual's private life in a secret register and the release of such information amounted to an interference with his right to respect for private life.[20] In 2000, it ruled that the very act of storing personal information by public authorities interfered with an individual's right to privacy, and therefore it must be justified.[21] In a more recent case involving the Internet, the Court ruled that in order to guarantee individuals' rights to privacy, governments have the duty to implement laws reconciling the confidentiality of Internet services with the prevention of crime and the protection of fundamental rights.[22] Both the Directive and proposed Regulation reference privacy rights set out in the Charter and the Convention.

The European approach to privacy often conflicts with the American "piecemeal" framework of torts, legislation, and weak constitutional guarantees, all of which are severely constricted by the First Amendment. According to Whitman, the United States "is much more oriented toward values of liberty, and especially liberty against the state" and comparatively much less concerned about the media.[23] Certainly, tensions exist in U.S. law regarding balancing rights to privacy with rights to free expression, but the Supreme Court's interpretation of the First Amendment is often squarely at odds with the European emphasis on privacy rights. This is not to suggest that Americans are unconcerned about privacy rights, particularly in the online environment. Although not the subject of this chapter, it is important to note that in response to U.S. citizens' concerns, the Obama administration has issued a Consumer Privacy Bill of Rights.[24] As new technologies proliferate, it is clear that privacy concerns will continue to escalate on both sides of the Atlantic.

EUROPEAN DATA PROTECTION DIRECTIVE AND PROPOSED REGULATION

The EU Data Protection Directive, which is at the center of the European data protection framework, was passed in 1995 by the EU's two main bodies—the

European Parliament and the Council of Europe. The Directive required EU member states to pass national legislation implementing the Directive's provisions by 1998.

The Directive seeks to balance rights of information privacy with advantages derived from the free flow of data. It applies to any "processing of personal data wholly or partly by automatic means" as well as some manual processing and applies to processing done by the private and public sector. The terms used in the Directive often are broadly defined. For example, processing by automatic means has been interpreted as applying to the dissemination of information via personal web pages.[25] The term "personal data" has been interpreted as including information in public records as well as information that already has been made public.[26]

Those determining the "purposes and means of processing the personal data" are designated as "data controllers" and have specific legal responsibilities. Each EU member country is required to create a supervisory authority that is charged with ensuring that data controllers comply with the national data protection law implementing the Directive. Numerous U.S. corporations have also voluntarily signed onto the EU/U.S.-negotiated "Safe Harbor Principles," issued by the Department of Commerce after the EU Directive entered into force. The Safe Harbor Agreement stipulates that U.S. companies that are active in the EU market must comply with EU privacy rules even when their data are processed in the United States.[27]

The Directive provides rights for individuals to gain access to information about the data that is held and processed about them.[28] If the processing does not comply with the Directive, particularly because personal information is inaccurate or incomplete, individuals have the right to have it corrected, restricted or even erased.[29] This is similar to the U.S. Privacy Act of 1974, enacted following the uproar over former President Nixon's actions, which guarantees citizens the right to see what information the government is collecting about themselves.

The proposed Data Protection Regulation builds on the current Directive but significantly bolsters these rights to provide more protection and enforcement in the digital environment. According to the explanatory notes of the proposed regulations, "Rapid technological developments have brought new challenges for the protection of personal data. The scale of data sharing and collecting has increased dramatically. Technology allows both private companies and public authorities to make use of personal data on an unprecedented scale in order to pursue their activities. Individuals increasingly make personal information available publicly and globally. Technology has transformed both the economy and social life."[30]

The proposed Regulation, which is expected to be adopted by late 2014, requires data controllers to provide much more detailed information to individuals, including how long their data will be stored, the data subject's rights, and where the data came from.[31] The Regulation also offers much stronger rights for the correction and erasure of data, including the "right to be forgotten," discussed below.

Although the Directive and proposed Regulation do provide exemptions for some or all of its provisions, the exemptions based on the protection of fundamental rights, particularly the right to freedom of expression, raise several concerns about whether information privacy rights are being appropriately balanced with rights to free expression and government transparency.

BALANCING FREEDOM OF EXPRESSION AND INFORMATION PRIVACY

The Court of Justice of the European Union (ECJ) has held that implementation of the Directive cannot run counter to the fundamental rights guaranteed by the European Charter and European Convention of Human Rights.[32] Included in both instruments is the right to freedom of expression, which includes the right to receive and impart information.

The Directive and proposed Regulation provide exemptions for data processing "solely for journalistic purposes or the purpose of artistic or literary expression" in order "to reconcile the right to privacy with the rules governing freedom of expression."[33] But it is up to each country to pass legislation clarifying how this is to be done.[34]

Media outlets have not been happy regarding how member states have balanced these rights. The European Newspaper Publishers' Association, which represents more than 3,000 news outlets across 21 European countries, stated that the Directive, as implemented in UK law, has had a negative effect on newsgathering and reporting because it "has decreased the flow of information to the public. The police and other public authorities have relied upon the data protection principles as an excuse not to make public information that was previously made publicly available or passed on to the press."[35] The European Federation of Journalists also expressed concern stating that a "better balance between freedom of expression and information and privacy rights is urgently needed."[36]

Just what article 9, the Directive's journalistic exemption, entails remains unclear. As of this writing, the ECJ has only addressed this exemption in two decisions. The first case involved a Swedish woman, Bodil Lindqvist, who faced criminal charges for uploading personal information about parishioners at her church. Lindqvist voluntarily set up her website to provide fellow parishioners information about their parish. She posted information about some parishioners including names, phone numbers, and employment information, and revealed that one parishioner was on half-time medical leave due to an injured foot. Lindqvist's postings were uploaded without informing the parishioners, getting their consent, or informing the Swedish data protection authority, which is required by law when processing personal data. After learning that some parishioners were upset, she promptly removed her website from the Internet. Nonetheless, a public prosecutor brought charges against her under Swedish legislation implementing the Directive.[37]

The second case involved the dissemination of individual tax income data of over one million Finnish citizens that were retrieved from public government documents. The company Satamedia published the information in a newspaper and in an electronic format that provided a pay-per-view text-messaging service. Although the newspaper provided some articles and advertisements, its main purpose was to provide income tax information about wealthy citizens. The subjects of the tax information were not informed before publication, but did have the opportunity to have their information removed upon request. After several complaints were filed, the Finnish data protection authority ordered publication of the tax data to cease.[38]

In both cases, the ECJ ruled that the disclosure of personal information did entail "processing of personal data wholly or partly by automatic means" within the meaning of the Directive,[39] so the question remained whether these activities fell under the Directive's journalistic exception. In Lindqvist, the ECJ punted the question back to EU member states. Acknowledging that Lindqvist had the right to freedom of expression, the Court stressed that governments have some discretion—or "margin to manoeuvre"—when implementing the Directive. Nonetheless, according to the Court, the laws of member states must provide "predictability" and not be contrary to protected fundamental rights, such as freedom of expression.[40]

The ECJ provided a little more guidance in the Satamedia ruling, although many questions still remain. According to the Court, the Directive's journalist exemption not only applies to "media undertakings," but to anyone involved in journalism, which must be defined broadly due to the importance of freedom of expression in a democratic society. The Court further explained that the type of communication medium used and whether a profit was made are not determining factors when interpreting the journalistic exemption. Nonetheless, the Court emphasized that any limitations on the protection of personal data must apply only as "strictly necessary."[41] The Court did clarify that Satamedia's activities "relating to data from documents which are in the public domain under national legislation must be considered as activities involving the processing of personal data carried out 'solely for journalistic purposes,' within the meaning of [the exemption], if the sole object of those activities is the disclosure to the public of information, opinions or ideas. Whether that is the case is a matter for the national court to determine."[42]

Based on the Satamedia ruling, the proposed Regulation provides much more detail in its recitals (explanatory notes contained within the preamble) than those of the Directive.[43] It explains that the exemption applies particularly in regards to "audiovisual field and in news archives and press libraries" and states that "in order to take account of the importance of the right to freedom of expression in every democratic society, it is necessary to interpret notions relating to that freedom, such as journalism, broadly." Therefore, when creating exemptions to the regulation, member states "should classify activities as 'journalistic' ... if the object of these activities is the disclosure to the public of information, opinions or ideas" regardless of the medium used and regardless of whether the endeavor is for profit or not. The recital also clarified that exemptions should not be limited to just media organizations.[44]

Several questions are left unanswered by Lindqvist, Satamedia and the proposed Regulation's recital regarding how national authorities and courts should interpret the journalist exemption. For example, how should lawmakers and courts determined whether the "sole object" of an expressive activity is "disclosure to the public of information, opinions or ideas"? After the Satamedia case was returned to a Finnish national court, the court ruled that the journalistic exemption did not pertain to the defendant's activities. According to the court, the text-messaging service was not disclosing information "to the public," only to individuals requesting the information. It also ruled that the newspaper's unedited list of names was not journalistic in nature. Thus, under the Finnish data protection law that implements the Directive, the defendants were not protected by exemption and were ordered to permanently cease publication.

BALANCING TRANSPARENCY AND DATA PRIVACY

A general right of access to government records is not a recognised fundamental right in the European context,[45] although recent case law of the European Court of Human Rights suggests that it could be moving in that direction.[46] Nonetheless, the principle of transparency is recognized in the treaties establishing European Union,[47] and rulings of the ECJ often refer to the "principles of transparency" and transparency as an "objective of general interest recognised by the European Union."[48]

The Directive and proposed Regulation do not provide a specific exemption pertaining to official documents or principles of transparency, and the ECJ has made clear that disclosure of public records containing personal information is considered processing personal information for the purposes of the Directive.[49] Nonetheless, the recitals of the Directive and proposed Regulation state that national legislation can take the "principle of public access to official documents" into account when implementing data protection provisions.[50] Although the principle of transparency does not have to be balanced with the same consideration as recognized fundamental freedoms, such as freedom of expression, a balancing still needs to occur.

According to an opinion of Article 29 Working Party, an independent advisory body established by the Directive, the goal for greater transparency in the public sector "sits uneasily with the global dissemination" of personal data. To emphasize this point, the opinion discussed the French government's practice of excluding the names of individuals who gave up their original nationality from the online version of their official records. This practice is meant to protect these individuals from retaliation from those in their country of origin.[51]

The ECJ has held that national laws requiring disclosure of personal information in the name of transparency do not automatically violate the Directive as long as the laws are necessary and appropriate for the stated objective.[52] The case addressed an Austrian law requiring public bodies to disclose salaries and pensions of their higher-paid employees. The Court left it to the national courts to decide if the disclosure of salary information was truly necessary to ensure proper management of public funds.

The Court looked at a similar issue in a case concerning a federal law requiring Internet publication of names of those receiving government agricultural funds. In determining if the interference with the individual's privacy rights was justified, the Court held that the law met the objective of general interest recognized by the EU to enhance transparency, but that it was not proportional. According to the Court, there were other means for achieving transparency that were far less intrusive.[53]

The above cases illustrate the ongoing issues regarding balancing the right to information privacy with rights to free expression and transparency, particularly in the context of e-governance. Implementation of a new "right to be forgotten" will likely intensify these issues.

THE RIGHT TO BE FORGOTTEN

The concept of a "right to be forgotten" evolved from the French and Italian legal doctrine of a "right to oblivion," which had been around since the 1970s. It refers to the right to not have information from your past disclosed in the present, and is understood mostly in the context of convicted criminals who have served their time, been rehabilitated or exonerated.[54]

The inclusion of a "right to be forgotten" in the proposed Regulation has sparked intense debate in part because it is unclear how a "right to be forgotten" would be implemented and interpreted in the digital age, particularly in the context of rights to free expression and government transparency. Part of the controversy stems from the fact there is no universal definition of a "right to be forgotten."

According to Bert-Jaap Koops, the "right to be forgotten" has been understood as three separate but overlapping concepts. In the most common articulation, it is understood as a right to have data about oneself deleted when it is inaccurate or no longer relevant. In the online context, this can include the right of individuals to have content that they uploaded about themselves removed. More controversially, it can include the right to have content about them removed that was uploaded by others.[55] A second articulation of the "right to be forgotten" focuses on the importance of a "clean slate," which stems in part from the "right to oblivion."

This is the idea that individuals and society benefit when "outdated negative information" about an individual is not allowed to be used against them. Koop explains that, since people and circumstances can change, "some areas of social life function better if people are given the chance to start from scratch."[56] This is illustrated by a recent case against Wikipedia by two German men who were convicted of murdering a famous German actor. The two men asked that their names be removed from the German Wikipedia website. Under Germany privacy law, the names of criminals who have served their time can be censored from news accounts. This case also highlights issues regarding implementation of a "right to be forgotten" in an online global environment. The German men also demanded removal of their names from the English-language version of the website.[57] Wikimedia, the parent company, was able to resist the order, but how such requests will play out under the European data protection framework remains unclear.

The third articulation also focuses on the importance of a "clean slate" but emphasizes individual self-development. According to this understanding, in order for individuals to be autonomous and freely express themselves, they need the "right to forget." In other words, individuals need freedom from the past in order to reinvent themselves in the present.[58] Mayer-Schönberger argues that in the age of Big Data, there needs to be a better balance between memory and forgetting. According to Mayer-Schönberger, as "more information is added to digital memory, digital remembering confuses human decision-making by overloading us with information that we are better off to have forgotten."[59]

Unfortunately, the draft Regulation does not provide definitional clarity regarding the "right to be forgotten." In fact, the term is not even defined, nor used in the wording of the provisions beyond its title, "The Right to be Forgotten and to Erasure." And, adding to the confusion, Vivane Reding, the European Data Protection Supervisor, stated that the "right to be forgotten" as proposed in the draft Regulation only strengthens rights that are already in the Directive.

Whether the draft Regulation creates a "new" right or not, it is clear that it is far more expansive and detailed regarding rights of access to personal information, rectification, restricting (referred to as "blocking" in the Directive), and erasure of personal data. Under the Directive, the right of erasure is confined to the rectification of inaccurate or incomplete data or when an individual can prove "compelling legitimate grounds." The draft Regulation makes it much easier for individuals to demand erasure of their personal data including data uploaded about them by others. In other words, information that has already been made public, even if it came from official records, could be considered private and removed.

The draft regulation also requires that the controller of the data in question notify "each recipient to whom the data have been disclosed."[60] In fact, a report by an independent advisory body established by the EU Data Directive, argues that the proposed Regulation should require all recipients to erase the data.[61]

The proposed "right to be forgotten" raises serious questions regarding freedom of expression and government transparency. As discussed above, the draft Regulation states that "journalist purposes" should be defined broadly, but how this will be interpreted and implemented in national law remains unclear. The draft Regulation recitals states that extra care should be taken regarding the "audiovisual field," "news archives," and "press libraries," but these are not defined.

Implementation of the "right to be forgotten" could also impact the public's ability to receive information from official documents. As discussed above, there is no exemption regarding public records. How will a "right to be forgotten" affect e-government initiatives? What about non-governmental organizations that gather and provide access to government documents? And what about search engines, such as Google, that provide links to webpages including those providing access to official documents? Can they be forced to withhold websites from their search results? The Google case discussed in the introduction illustrates these concerns.

Whether the "right to be forgotten" in the proposed regulation is a "new" right remains to be seen. Regardless, the impact of stronger rights of erasure and restriction will come down to implementation and enforcement. According to Erdos, enforcement of the Directive has been weak. "The basic, albeit unpalatable, truth is that from the beginning this body of law has been routinely and often necessarily ignored, misapplied, and/or evaded."[62]

The proposed Regulation substantially bolsters enforcement including mandatory fines up to 500,000 Euros or up to 1 percent of a company's worth, up to one million euros. How the "new" provisions are implemented will be influenced by new enforcement measures, and the inclusion of a "right to be forgotten," whether new or not, may empower governments to strengthen these rights beyond what is required by the Regulation. If this is the case, much greater clarity will be needed regarding interpretation of the journalistic exemption.

CONCLUSION

As "Big Data" continue to grow and evolve so will calls for privacy protection. Ironically, the same technologies that we fear provide us with ever greater opportunities to exercise our rights of freedom of expression and access to information. These are not simply European issues. Although the U.S. and European understanding of privacy differ, it would be wrong-headed to assume that elements of a European framework of privacy, particularly in the online context, could never develop in the U.S. According to Steven Bennett, "Despite cultural divisions between the European Union and the United States on the substance of privacy rights and the reach of jurisdiction over Internet-related activities, a process of 'convergence' in views seems almost inevitable."[63] Ultimately, what is needed is more expression—debate on a global scale. Our digital future depends on it.

NOTES

1. Suzanne Daley, "On Its Own, Europe Backs Web Privacy Fights," *New York Times*, August 9, 2011, A1.
2. "Google Fights Spanish Privacy Order in Court," *The Guardian*, January 20, 2011.
3. David Banisar, "Linking ICTs, the Right to Privacy, Freedom of Expression and Access to Information," *East African Journal of Peace & Human Rights* 16 (June 1, 2010): 124–154.
4. Josh Halliday, "Europe's Highest Court to Rule on Google Privacy Battle in Spain," *The Guardian*, March 1, 2011.
5. Cheryl Ann Bishop, *Access to Information as a Human Right* (El Paso, TX: LFB Scholarly Publishing, 2012).
6. Omer Tene and Jules Polonetsky, "Privacy in the Age of Big Data: A Time for Big Decisions," *Stanford Law Review Online*, February 2, 2012, http://www.stanfordlawreview.org/online/privacy-paradox/big-data.
7. Cisco® Visual Networking Index Initiative, *The Zettabyte Era*, May 30, 2012, http://www.cisco.com/en/US/solutions/collateral/ns341/ns525/ns537/ns705/ns827/VNI_Hyperconnectivity_WP.pdf.
8. Tene and Polonetsky, "Privacy in the Age of Big Data," 63.
9. David Banisar, "The Right to Information and Privacy: Balancing Rights and Managing Conflicts," *World Bank Institute Governance Working Paper* (March 10, 2011): 8, http://ssrn.com/abstract=1786473.
10. Directive 95/46/EC, "On the Protection of Individuals with Regard to the Processing of Personal Data and on the Free Movement of Such Data," European Parliament and the Council of the European Union (October 24, 1995), http://eur-lex.europa.eu/LexUriServ/LexUriServ.do?uri=CELEX:31995L0046:en:HTML.
11. Jack Goldsmith and Tim Wu, *Who Controls the Internet: Illusions of a Borderless World* (New York, Oxford University Press, 2006): 174.
12. European Commission, "Proposal for Regulation of the European Parliament and the Council of the European Union on the Protection of Individuals with Regard to the Processing of Personal Data and on the Free Movement of Such Data (General Data Protection Regulation)" (January 25, 2012), http://www.europarl.europa.eu/registre/docs_autres_institutions/commission_europeenne/com/2012/0011/COM_COM(2012)0011_EN.pdf.
13. Adam Thierer, "Erasing Our Past on the Internet," *Forbes*, April 17, 2011.
14. Peter Fleischer, "Hey, Mom and Dad, Look, I'm the Most Powerful Censor on the Internet," *Peter Fleischer: Privacy?...*" (February 15, 2012), http://peterfleischer.blogspot.com/2012_02_01_archive.html.
15. "Convention for the Protection of Human Rights and Fundamental Freedoms," (November, 4, 1950), Art. 8, http://conventions.coe.int/Treaty/en/Treaties/Html/005.htm. The Convention is more commonly referred to as the European Convention on Human Rights.
16. James Q. Whitman, "The Two Western Cultures of Privacy: Dignity Versus Liberty," *Yale Law Journal* 113 (April 2004): 1161–1221.

17. Lauren Movius and Nathalie Krup, "U.S. and EU Privacy Policy: Comparison of Regulatory Approaches," *International Journal of Communication* 3 (2009): 172.
18. Charter of Fundamental Rights of the European Union, December 7, 2000, http://www.europarl.europa.eu/charter/default_en.htm
19. European Agency for Fundamental Rights, *Data protection in the European Union: The Role of National Data Protection Authorities* (Luxembourg: Publications Office of the European Union, 2010): 6, http://fra.europa.eu/sites/default/files/fra_uploads/815-Data-protection_en.pdf
20. *Leander v. Sweden*, European Court of Human Rights, March 26, 1987, para. 48, http://hudoc.echr.coe.int/sites/eng/pages/search.aspx?i=001-57519
21. *Amann v. Switzerland*, European Court of Human Rights, February 16, 2000, para 70, http://hudoc.echr.coe.int/sites/eng/pages/search.aspx?i=001-58497
22. See, for example, *K.U. v Finland*, European Court of Human Rights, February 3, 2009, http://hudoc.echr.coe.int/sites/eng/pages/search.aspx?i=001-89964
23. Whitman, "The Two Western Cultures of Privacy," 1162.
24. The White House, *Consumer Data Privacy in a Networked World: A Framework for Protecting Privacy and Promoting Innovation in the Global Digital Economy* (February 2012), http://www.whitehouse.gov/sites/default/files/privacy-final.pdf
25. Criminal proceedings against Bodil Lindqvist, Court of Justice of the European Union, November 6, 2003, http://curia.europa.eu/juris/celex.jsf?celex=62001CJ0101&lang1=en&type=NOT&ancre=
26. "Public Sector Information and the Protection of Personal Data," Article 29 Working Party Opinion 3/99 (May 3, 1999), 3–4.
27. See Safe Harbor, Export.gov, http://export.gov/safeharbor/; see also Working Party on the Protection of Individuals with Regard to the Processing of Personal Data, Opinion 7/99 On the Level of Data Protection Provided by the "Safe Harbor" Principles, Doc. No. 5146/99/EN/final (Dec. 3, 1999).
28. EU Data Protection Directive, Art. 12(a).
29. EU Data Protection Directive, Art. 12(b).
30. EC Proposed Data Protection Regulation, explanatory memorandum 1.
31. EC Proposed Data Protection Regulation, Arts. 14 and 15.
32. *ERT* v. *DEP* (June 18, 1991), para. 44, http://curia.europa.eu/juris/celex.jsf?celex=61989CJ0260&lang1=en&type=NOT&ancre=
33. EU Data Protection Directive, Art. 9; EC Proposed Data Protection Regulation, Art. 80(1). Article 9 of the Data Protection Directive limited the exemption by the phrasing "only if they are necessary" to reconcile the competing rights. According to James Maxeiner, the phrase was added after the U.K. Data Protection Registrar expressed concern that balancing free expression and privacy might favor the media too much, James R. Maxeiner, "Freedom of Information and the EU Data Protection Directive," 48 Fed. Comm. L.J. 93, (1995), 102. The EC proposed regulation changed the phrasing to "in order to."
34. EU Data Protection Directive, Recital 37. The EC proposed regulation requires member states to notify the European Commission of the provisions of the adopted law, EC Proposed Data Protection Regulation, Art. 80(2).
35. European Newspaper Publishers' Association, "Review of Data Protection Directive" (Aug. 1, 2002), http://ec.europa.eu/justice/policies/privacy/docs/lawreport/paper/enpa_en.pdf
36. European Federation of Journalists, "Contributions to the Questionnaire for Stakeholders Consultation on Data Protection," (July 1, 2010), http://europe.ifj.org/assets/docs/194/151/eff22c2-4b28997.pdf

37. Bodil Lindqvist, para. 12–15.
38. Satakunnan Markkinapörssi and Satamedia (Dec. 16, 2008), para. 25–34, http://curia.europa.eu/juris/celex.jsf?celex=62007CJ0073&lang1=en&type=NOT&ancre=
39. Bodil Lindqvist, para. 26–27; Satamedia, para. 37.
40. Bodil Lindqvist, para. 84–87.
41. Satamedia, para 54–62.
42. Satamedia, para. 62.
43. EC Proposed Data Protection Regulation, at 3.4.9 of the Explanatory Memorandum.
44. EC Proposed Data Protection Regulation, Recital 121.
45. Sacha Prechal and de Leeuw, Magdalena, "Transparency: A General Principle of EU Law?," in *General Principles of EC Law in a Process of Development*, European Monographs, Number 62, eds. Ulf Bernitz, Joakim Nergelius, and Cecilia Cardner (The Hague: Kluwer Law International, 2008), 210.
46. Bishop, *Access to Information*, 42–45.
47. Treaty on European Union, Art. 11 and the Treaty on the Functioning of the European Union, Art. 15, as amended by the Treaty of Lisbon (Dec. 1, 2009), http://eur-lex.europa.eu/JOHtml.do?uri=OJ:C:2010:083:SOM:EN:HTML
48. See, for example, Volker und Markus Schecke, European Court of Justice (Nov. 9, 2010), para. 68, http://curia.europa.eu/juris/celex.jsf?celex=62009CJ0092&lang1=en&type=NOT&ancre=
49. Österreichischer Rundfunk and Others, European Court of Justice (May 20, 2003), http://curia.europa.eu/juris/celex.jsf?celex=62000CJ0465&lang1=en&type=NOT&ancre=
50. EC Proposed Data Protection Regulation, Art. 18; EU Data Protection Directive, Art. 17.
51. Opinion No. 3/99, p. 7.
52. Rundfunk and Others, European Court of Justice, para. 94.
53. Schecke, European Court of Justice, para. 81–3.
54. Paul A. Bernal, "A Right to Delete?" *European Journal of Law and Technology* 2 (2, 2011), sec. 1.1, http://ejlt.org//article/view/75/; Jeffery Rosen, "The Right to be Forgotten," *Stanford Law Review Online* (February 13, 2012), 88, http://www.stanfordlawreview.org/online/privacy-paradox/right-to-be-forgotten
55. Bert-Jaap Koops, "Forgetting Footprints, Shunning Shadows: A Critical Analysis of the 'Right to be Forgotten' in Big Data Practice," *SCRIPTed* 8 (December 2011), 237.
56. Koops, "Forgetting Footprints," 250.
57. John Schartz, "Two German Killers Demanding Anonymity Sue Wikipedia's Parent," *New York Times*, November 13, 2009.
58. Koops, "Forgetting Footprints," 252–253.
59. Viktor Mayer-Schönberger, *Delete: The Virtue of Forgetting in the Digital Age* (Princeton University Press, 2011): 163–164.
60. EC Proposed Data Protection Regulation, Art. 13.
61. "On the Data Protection Reform Proposals," Article 29 Working Party Opinion 01/2012 (March 23, 2012), 13–14, http://ec.europa.eu/justice/data-protection/article-29/documentation/opinion-recommendation/files/2012/wp191_en.pdf, 13–14, http://ec.europa.eu/justice/data-protection/article-29/documentation/opinion-recommendation/files/2012/wp191_en.pdf
62. David Erdos, "The Tension between Data Protection and Freedom of Expression," *Eurozine* (May 6, 2011), 2, http://www.eurozine.com/pdf/2012-06-05-erdos-en.pdf
63. Steven Bennett, "The 'Right to Be Forgotten': Reconciling EU and US Perspectives," 30 *Berkeley Journal of International Law* (1, 2012): 161, 192–93.

PART TWO

Online Dilemmas: Email, Social Media, and Those Pesky Jammie Surfers

CHAPTER FIVE

Electronic Court Record Access: Present Landscape, Neutral Principles, and the Looming Interloper of Contextual Privacy

RICHARD J. PELTZ-STEELE

Across centuries of common-law development, access to the judiciary has foundered on the shoals of star chambers and secret dockets. But the courts also have safeguarded accountability and direct democracy by championing ideals such as witness confrontation, the jury trial, and the open courtroom. The U.S. judicial system navigated the civil rights era without the supercharge of a freedom of information act, partly because the courts, of all government forums, were the least secretive and the least hostile to challengers of the status quo amid executive scandals and foreign wars. For the ordinary citizen, inspecting a record at the window of the clerk of court was much the same affair in 1960 and 1990 as in 1790 and 1860.

Then came the information age. With the unprecedented capacity of electronics to gather, manipulate, and disseminate information, the game changed, and the courts are no longer above the fray. In the 1990s, policymakers turned serious attention to court record access in state and federal governments. Court record access in the generation since has begun a revolution on par with the move in executive record access from the 1946 Administrative Procedure Act[1] to the 1996 Electronic Freedom of Information Act Amendments.[2] Law and policy continue to develop apace with technology, and many conflicts familiar to access and privacy advocates in the state and federal FOIA arenas are playing out in parallel court chronicles.

This chapter examines the current landscape in court record access policy with particular attention to key issues in present development, particularly online access. Judges have grappled with the concept that citizens in many parts of the country can now sit at home in their pajamas and peruse online court files about their neighbors, friends, relatives, and co-workers. The chapter also explains the deliberate and central role of neutrality in FOI policy since the civil rights era and how neutrality is jeopardized in court record access policy development. Then, this chapter briefly describes the emergence of context as a central principle in recently influential thinking about privacy. The chapter concludes that incompatibility between neutral principles and new privacy paradigms jeopardizes transparency and accountability in the courts.

THE PRESENT LANDSCAPE OF ELECTRONIC COURT RECORD ACCESS

Courts were largely immune to the statutory access revolution of the 1960s to afford citizen access to executive branch records in the states and federal government. Even when state FOIAs purported to apply to courts, the scope was limited functionally to court administration and yielded to narrow statutory directives and case-specific exercises of inherent judicial powers.[3] Chiefly, the courts continued to apply a common law framework to record access. Twentieth-century common law was liberated from old British shackles—an origin shared with statutory systems—such as the litigation, or special, interest rule, which sharply limited access to directly concerned requesters.[4] But the rebuttable presumption of access that characterizes modern common law continued to employ value-driven inquiries, essentially balancing tests, permitting consideration of factors such as requester purpose.[5] The common law approach provided cover for broad official discretion, which wrought unpredictability and divergent analyses across jurisdictions and judges.[6]

Two forces coalesced in the information age to force the courts in the 1990s to confront access policy questions in a systematic fashion.[7] First, the migration of court records to electronic platforms meant that vast stores of data were newly available for compilation, analysis, and redistribution with unprecedented efficiency. Records moved from the purview of local judges and clerks to the computerized stores of court administrators. Data brokers, researchers, lawyers, and journalists were among interested readers who converged on these resources with motives from profit to public interest. Second, and consequently, public anxieties over personal privacy surged with the proliferation of electronic record management. The courts were not spared the same public demands for thoughtful policy development that resulted in a wave of electronic amendments to state and federal access laws.[8]

The federal courts' PACER (Public Access to Court Electronic Records) system went live in 1988 as a limited convenience for professionals and libraries.[9] But the rapid rise of networked home computing meant that PACER and its state analogs became the principal channels of public access to the judiciary.[10] After examination study in the 1990s, the Administrative Office of the U.S. Courts in 2001 began rollout of a nationwide web-based case management system with the bankruptcy courts,[11] and the Judicial Conference of the United States adopted an access and privacy policy later that year.[12] In 2002, the Conference of Chief Justices (CCJ) and Conference of State Court Administrators (COSCA) approved model rules to guide state policy development.[13] Other models followed, but the CCJ/COSCA guidelines were especially influential as at least 23 states studied court record access from 2002 to 2005.[14]

In the present landscape, about half of the states have electronic case record access policies.[15] The development of these policies remains highly fluid, and their terms highly variable. In 2011, four new or revamped state policies went into effect—in Hawaii, Missouri, New Jersey, and Wyoming—and a first ever policy suspension occurred in Montana amid a controversy over electronic access to family law case records.[16] California and New York offer examples of access-restrictive and access-friendly approaches, respectively, though New York operates under the guidance of an access-friendly study and no comprehensive rule.[17] The federal system, through PACER, continues to model an approach closer to the access-friendly side of the spectrum. In keeping with procedural trend, the federal system—by effect of the E-Government Act of 2002[18] and consequent promulgation of federal rules[19]—commands attorneys to redact limited confidential information from the public record upon filing, then regards the rest as open.[20]

Work in the last decade was critical to define terms and identify issues. Definitionally, court policies tend to distinguish administrative records from case records.[21] Administrative records may remain subject to public disclosure on FOIA terms or be subject to a separate but more FOIA-like access regime than governs case records. Case records are the heart of the matter in court record access. Typically, at minimum, they are the records upon which a court makes its decisions.[22] Excluded are, for example, the fruits of discovery that are not filed with the court.[23] A more limited-access approach might distinguish "court records" authored by the court from records submitted by litigants to the "case file."[24]

The CCJ/COSCA guidelines crystallized the issues policymakers would face and choices they could make.[25] A range of experiments unfolded in the states. Following the example of common law and the federal system, court record access policies begin with public access as presumptive. That presumption may be inverted or rebutted upon an array of countervailing priorities.[26] These priorities account for roughly three distinctions in case records: (1) access at the courthouse or from a remote electronic terminal; (2) access to individual case records or to multiple records, whether in bulk or compiled upon a search; and (3) access to records or information within or not within categories akin to conventional FOIA exemptions.

REMOTE ONLINE ACCESS AND JAMMIE SURFERS

Much angst has accompanied the advent of court record access by the general public through electronic portals from remote locations.[27] The essence of the concern arises from the lack of a human intermediary with discretion.[28] Well-founded fears surround the possibility of a violent offender learning the address of a victim, or organized crime foretelling a police search. "Whosarat.com" has been a flashpoint for debate, as the website uses information from sources including public records to unmask informants and undercover agents.[29] More nebulous concerns arise from the "jammie surfer" problem: the pajama-clad voyeur who reads about her neighbors' personal lives from the privacy of her home.[30] Somewhere in between are the television reporter pursuing the details of a parent's drug addiction in a child-neglect case[31] and the gossipy teen stumbling upon adultery allegations in the divorce of a friend's parents.[32] The geographically indiscriminate reach of the Internet also complicates remote access with a problem termed "hyperdissemination."[33] According to one anecdote, for example, the details of a woman's divorce in an American court became common gossip in her home community in India, where a powerful social stigma attached.[34]

Remote access is the raison d'être of PACER. But remote access is moderated in that PACER—in contrast with state systems such as California's[35]—requires user registration and tracks activity. So if a particular datum gleaned from a not-oft-viewed court record is later employed, perhaps to intimidate a witness,[36] it might be possible to track down the virtual record requester.[37] The Judicial Conference was abundantly cautious in moving criminal case files onto PACER, but did so after nailing down a list of exemptions, including juvenile records and juror identifications, for both computer and courthouse.[38] At the state level, the Oklahoma Supreme Court fractured in 2008 over online publication and "light switch[ed]" access off, then on.[39] Missouri typifies a reserved approach, limiting online access to enumerated case-biographical data.[40]

Classes of cases, such as family, juvenile, and criminal, are a common place to draw the line on remote access. New York policy does not discriminate on the medium or remote location of access.[41] In contrast, California restricts record access to family and criminal matters to the courthouse only[42]—artificially effecting practical obscurity, which Dooley rechristened "intentional inconvenience"[43]—though a special rule liberalizes access in "extraordinary criminal cases" and remote access in principle is "encourage[d]."[44] PACER, though liberal on civil cases generally, limits social security and immigration appeal records to the courthouse.[45] Minnesota courts prohibit all electronic access to child protection records, even when a record may be disclosed in hardcopy, thus effecting practical obscurity through medium distinction.[46] New Jersey courts are just beginning to explore whether to take a location-neutral approach to the online dissemination of their partly public family law records under a rule that is silent on the question.[47]

BULK AND COMPILED ACCESS

Bulk and compiled electronic record disclosures have long aggravated access and privacy tensions and more recently, as information has been commodified, have divided access advocates in journalism and business.[48] In federal FOIA construction, the U.S. Supreme Court in 1989 famously, or infamously, distinguished criminal histories available in the "practical obscurity"[49] of paper files in the nation's vastly dispersed local courthouses from access to the same information compiled in a nationwide law enforcement database.[50] Moreover, in the decades since, businesses such as data brokers, banks, and retailers have become expert at assembling from disparate sources detailed profiles of individuals, adapting to daily life the "mosaic" approach in foreign intelligence gathering.[51] So profound has been the tide of public hostility to Orwellian information flows and risks such as identity theft[52] that access advocates in journalism at last distanced themselves from commercial interests. The Supreme Court upheld a distinction in California FOI law that afforded noncommercial requesters of police records preference over commercial requesters.[53] Information commodification, moreover, raises questions besides privacy, such as whether record system costs should be borne by requesters rather than taxpayers and how the courts can maintain control over information after its distribution to effect expungement and error correction.[54] Related problems arise in the privatization of record management;[55] the practical and financial terms of bulk and compiled public access should be contemplated in advance of outsourcing.[56]

The federal system struggles with the myriad problems of bulk and compiled access. PACER architecturally encumbers bulk or compiled data capture, a coincidence of design and desire that facilitates record revision, supports user registration, and sustains user-fee-based financing. Even when researchers have resources or tech savvy to download records en masse, per-page fees are prohibitive.[57] Debate erupted around and halted a fee-free pilot program when access activists were discovered expeditiously downloading the database.[58] Meanwhile, the Princeton-housed RECAP project is "turning PACER around."[59] RECAP offers to PACER users a free browser add-on with access to a file-sharing database, accomplishing for court records what copyright precluded for popular music. PACER use policies promise punishment for, inter alia, "[a]ny attempt to collect data from PACER in a manner which avoids billing," but PACER also expressly disclaims responsibility for downstream disseminations.[60]

Cognizant of the policy implications, authors of the CCJ/COSCA guidelines modeled disparate treatment for individual record access on the one hand and for bulk or compiled access on the other.[61] The latter requests may be restricted universally, or may be restricted upon a collateral policy trigger, such as remote access, commercial purpose, or class of record.[62] The California superior court database permits only case-by-case record access except for index data.[63] New York has not yet tackled bulk and compiled record requests, like many states, but continues to be guided by its medium-neutral philosophy and protected by a web architecture designed only for case-specific access.[64] Arkansas devised two tracks for bulk and compiled access that modestly abandon neutrality to distinguish noncommercial from commercial requesters.[65] Modeled on the state's neutrality-principled FOIA, one track affords generous access to noncommercial requesters including media. Alternatively, requesters must identify themselves and their purposes, justify their requests in the public interest, explain their protocols to ensure data security and accuracy, and submit to the discretion of court administrators, which may include discretionary fees and restrictions on subsequent dissemination.

CATEGORICAL EXEMPTIONS

Exemption from disclosure for categories of records or data has been the stuff of access debates in the legislatures since the 1960s. Policymakers in court record access have to decide whether and how (a) to accommodate existing legislative "carve-outs" from presumptive public disclosure, such as the veil of secrecy that ordinarily cloaks adoption records; (b) to preserve traditional judicial prerogatives, such as the inverse presumption for grand jury transcripts and pending search warrants, as well as trial court discretion to grant case-specific protective orders and seals; and (c) to create new exclusions based on values such as personal privacy and business competitiveness, whether for classes of records, such as domestic relations pleadings, or for sensitive information, such as financial account numbers, trade secrets, or the names of sexual assault victims.

Content-based exemptions from public access to court records have occupied the ground where common law and statutory court procedure meet. Court record access policies universally preserve inherent and rule-based judicial discretion, case by case, to rebut the common law access presumption with a narrowly tailored seal or "good cause" protective order—or rarely, as in the case of grand juries and unserved search warrants, to preserve secrecy upon an inverse common law presumption.[66] Traditional fodder for protection is, for example, a trade secret, which might be subject to litigation rendered moot by revelation.[67] Personal privacy of the everyday sort—that is, not the constitutional right implicated by *Griswold v. Connecticut*[68] or *Whalen v. Roe*[69]—does not ordinarily justify closure.[70] But having shifted the redaction burden to litigants in the first instance, court policies have taken to dictating both what may be filed under seal and, thus inversely, what may not be filed under seal without a court order, fueling an increase in requests for those orders. Studies have found an excess of sealing, including seals that outlived their apparent justifications.[71] The Judicial Conference responded modestly with a policy discouraging federal judges from sealing entire civil case files absent "extraordinary circumstances" and less restrictive alternatives.[72]

Court record access policies also preserve legislative exemption for information and records by category.[73] Traditional targets for these measures include juvenile criminal matters, domestic proceedings involving abuse or juvenile welfare, sexual assault victim identifications, and expunged dispositions. At least in the context of courtroom access, exclusions are constitutionally suspect.[74] The federal picture on constitutional record access is not yet clear,[75] but state constitutional rights also may be implicated.[76] Separation of powers issues also abound, though courts usually are game to follow the statutory lead when the legislature has assumed political responsibility.[77] Thus the New Jersey policy, for example, professes a general rule of fealty to "statute, rule, or prior case law," and accordingly enumerates dozens of exemptions with citations to preexisting authority.[78]

The fluctuating variable in court record access exemptions has been the broadening or creation of exemptions in court record access policies themselves. In state and federal policies, a narrow consensus has emerged around the exemption of mostly numeric identifiers: social security, taxpayer ID, and financial account numbers save the last four digits; birthdates save the year; and minors' names save initials.[79] Strategically, media access advocates might let these exemptions slide,[80] but otherwise argue vigorously that court policymakers lack constitutional or statutory authority to cook up new, case non-specific exemptions.[81] Privacy advocates meanwhile have recognized the judiciary as a new front in the access debate and locate the power to create exemptions within the scope of judicial rulemaking authority.[82]

The result of this tension in some jurisdictions is categorical exemptions that are contingent on a collateral policy trigger, such as remote access.[83] Thus the federal and California systems, as described previously, employ categorical triggers to restrict remote access to more than docket and disposition. Representatively of state and federal policies, New Jersey courts restrict access to numeric personal identifiers,[84] but assert no new, generalized privacy exemption beyond extant statutory and constitutional rights of record subjects.

Going forward, then, controversy in the ongoing substantive development of court record access policies surrounds remote access and bulk and compiled access, each with the added dimension of categorical triggers. Procedural issues, such as the mechanics of dual filing or filing under seal to protect confidential information, will continue to be problematic, especially amid record management privatization,[85] pro se filing increases,[86] and insatiable media appetite.[87] Amid an upswing in the vigor of privacy arguments, to be discussed below, there are broader issues on the horizon such as the possible redefinition of case records, which could limit or preclude access to pleadings and motions and force constitutional questions.[88] Short of that extreme, wrangling is occurring on the margins over access to records such as exhibits, motion attachments, and metadata.[89]

THE HAPPY MARRIAGE OF ACCESS AND NEUTRALITY

Neutrality as a controlling principle was a deliberate and central feature of the FOIA revolution. The historic special interest rule turned on the requester's articulation of identity and purpose, namely a legally protectable interest.[90] In the 20th century and still today when common law complements statute, public interest in disclosure became an element in presumption-and-rebuttal analysis.[91] The element sometimes drifted from the requester's subjective motive to an objective policy test.[92] The objective focus accords with statutory FOI, which typically is neutral in the subjective dimensions of requester identity and motive. Otherwise stated, FOI norms proclaim the immateriality of requester identity and motive, and access turns only upon the objective record.[93]

FOIAs came about specifically to ameliorate the inconsistencies and potential for abuse inherent in the discretionary determinations that subjectivity invited. Reflecting favorably on the open records laws of 27 states in 1953, Cross, who was instrumental in the creation of the federal FOIA, cited "citizen" and "any person" provisions as "intended to remove the common law necessity that the applicant show some sufficient 'special interest'"[94] and "that the applicant allege and prove a 'proper purpose.'"[95] The evil these statutes sought to avoid was the sort of judicial discretion known to Massachusetts common law, Cross explained, where the high court was "seldom allergic to barriers and ever preoccupied with the prerogatives and convenience of …public servants."[96]

Before the federal FOIA, the "proper and direct concern" test under the Administrative Procedure Act had rationalized denial of access to the names and salaries of postal workers.[97] Clearly articulated categorical exemptions in the FOIA superseded "the vague phrases 'good cause found,' 'in the public interest,' and 'internal management'"[98] with which federal officials had hidden controversial contracting votes and multimillion-dollar bids. A House report reasoned, "No Government employee at any level believes that the 'public interest' would be served by disclosure of his failures or wrongdoings, but citizens both in and out of Government can agree to restrictions on categories of information …"[99] Requester neutrality and politically negotiated exemptions thus replaced subjective scrutiny.

Electronic FOIA extended neutrality in the electronic dimensions of medium, format, and location. Seven years after the Supreme Court fretted over lost practical obscurity, the 1996 Electronic FOIA Amendments rejected medium and format as "not relevant" and vested the requester with discretion.[100] E-FOIA, moreover, affirmatively required web publication of frequently requested records. The federal FOIA has never demanded that requesters present themselves in person, and state FOIA audits are replete with reports of improper intimidation when they do.[101]

Open records laws in the states almost uniformly have affected neutrality in the conventional dimensions of requester identity and motive and in the electronic dimensions of location, medium, and format. The "citizen" or "any person" approach is now universal,[102] compromised rarely by excepting prisoners for asserted security concerns.[103] Purpose neutrality also is universal in the states with scattered exceptions,[104] usually couched as exceptions to exemption, such as for motor vehicle records,[105] and sometimes to charge commercial requesters higher fees or to impede mailing-list compilation.[106] Following the federal lead, the states have widely rejected disparate treatment for conventional and electronic access.[107]

Four decades of experience with FOI in state and federal executive branches have established the superiority of access based on a presumption of openness mitigated only by politically negotiated categorical exemptions.[108] Broad official discretion upon vague standards such as "public interest" has invited abusive scrutiny of requester and purpose or overreliance on practical obscurity. Amorphous values such as "privacy" are equally unworkable if discretion is not constrained by accountable legislators and stringent tests, faithfully applied, such as exemptions 6 and 7(C) of the 1966 federal FOIA.[109]

Neutral principles have proven essential to government transparency and accountability, but in court record access policy, they are back on the negotiating table, often without the safeguards of political process. Disparate access for electronic, remote, and multiple-record requests has compromised neutrality in the electronic dimensions. Thus far, these compromises have turned on preexisting record or information exemptions, tested historically or politically to mitigate deviation from neutrality. But as developing policies and digitization expand the potential for remote and multiple-record access, the appeal grows too of vague standards affording broad official discretion.

LOOMING INTERLOPER OF CONTEXTUAL PRIVACY

With little more than a century to develop in common law—and really only half that time upon serious effort—the right of privacy has suffered a horde of surveyors who would tame its stretches with sorting, labeling, and the induction of pragmatic rules. The struggle has been frustrated by human behavior, which demonstrates a problematic and recurring conundrum: the simultaneous desire to protect and publish private information. A common feature of privacy as manifested in constitutional, tort, and property law is Solove's "secrecy paradigm" that a secret divulged is no longer secret.[110] Yet the person who readily jots her cell number on a form to win a car later objects to a timeshare peddler's robocall. The Facebook user wants to share party pictures with hundreds of "friends," but not have them fall into an employer's hands.[111] Saliently, people publish to the government—that's the "of the people" government—information about their money and health, but want to maintain control over that information after it leaves their hands. The access community has been exasperated by demand for a legal regime that respects government transparency but accommodates this seemingly cake-and-eat-it-too approach to privacy.

Solove[112] and Nissenbaum[113] have emerged lately from the pack of privacy thinkers with new theories that surmount the conundrum in an effort to construct a legal framework. With a bow to context, Solove constructed a taxonomy of privacy in sixteen problems as a foundation to identify compensable harms. With regard to court records, for example, he described problems in "aggregation," "secondary use," and "increased accessibility."[114] All three are implicated when, for example, hyper-dissemination makes private information submitted in family court a means to identity theft. Nissenbaum articulated a "framework of contextual integrity"[115] to determine when privacy violations occur. The framework examines information flow to detect changes in social context, actors, information attributes, or transmission principles. Remote and bulk electronic access to court records, for example, signal significant changes to the status quo: jammie surfing requesters (actors), aggregated information (attributes), and online retrieval (transmission). Resultant harms therefore may be averted "by obfuscating some of the information in the records or by ... limited access control or authorization mechanisms."[116]

These approaches to privacy resolve the conundrum by understanding that expectations survive dissemination. The taxpayer discloses to the IRS, but not to the general public, and the law backs that expectation. The 1974 Privacy Act limits information sharing between agencies for secondary uses, and prompted a pilot's suit when aviation and social insurance officials compared notes.[117] In the U.S. Supreme Court's 2012 rejection of warrantless GPS tracking data in a Fourth Amendment challenge to criminal conviction, Justice Sotomayor, concurring, considered the use of public data to assemble mosaic profiles of individuals.[118] She concluded that "it may be necessary to reconsider the premise that an individual has no reasonable expectation of privacy in information voluntarily disclosed to third parties."[119]

It is tempting then to fete context as the savior of privacy: to chart legally cognizable interests on the shores of reasonable expectations in context. But context ultimately encroaches on neutral principles. When access is shaded by context, the requester's identity and purpose become material; the location and medium of dissemination matter. If contextually significant changes are the disappearance of practical obscurity and the advent of electronic dissemination, then policymakers are justified in effecting obscurity through "intentional inconvenience."

This is not to say that context and neutrality cannot coexist. Case-by-case access necessarily turns on context, as when a court balances interests over access to closed-case search warrant affidavits.[120] Access policy also may comprise categorical generalizations. For example, rape shield laws statutorily restrict access to information about sexual assault complainants.[121] The statutes account for the expectations with which a complainant reveals intensely personal information— that it will be used only insofar as a court determines essential for criminal justice, not to fuel gossip or social stigma—as well as the context in which that information might be republished harmfully—to infotainment bloggers or nosy neighbors. These rules suggest divergence from neutrality. But the former determination is case-specific, a court ruling susceptible of objection and appeal. The latter determination is narrow, the result of political negotiation by accountable actors.

Court record access policies are problematic when they transgress neutral principles and exceed these bounds of case specificity, political accountability, and narrow articulation. There is not, as yet, a movement to adopt, across the board, vague exemption standards such as "good cause," nor is there a movement to protect privacy with unbridled discretion. But the trend in the states is to protect privacy at minimum through court-made rules that transgress neutral principles—for example, by conditioning electronic access on the requester's location, or by conditioning multiple-record access on the requester's purpose. Intentional inconvenience implicates architecture, like law, as regulator.[122] A court-ordered ban on Internet publication transgresses location neutrality and is case non-specific, supra-political, and all-encompassing. Such a rule may be justified with reference to the contextual paradigm, but not the secrecy paradigm. Too wide an arc of divergence from FOI norms smacks of the patronizing regard for public access that predated the FOIA revolution.

CONCLUSION

Court record access and privacy policies are evolving rapidly to meet technological challenges. Policies to date have tended only to codify the mix of common law and statutory rules that governed previously. But developments in information architecture are compelling and empowering policymakers to examine the scope of categorical exemptions and of remote and multiple-record access. At the same time, changes in thinking about privacy are challenging the conventional paradigm that information must be either secret or public. New paradigms instead emphasize nuanced expectations and context. A tempting balance can be drawn that would transgress well established neutral access principles with respect to the conventional dimensions of requester purpose and motive, and to the electronic dimensions of location, medium, and format. This approach would reserve superior access rights for qualifying requesters, or obstruct with intentional inconveniences those with inferior rights, and thereby better comport with the contextual paradigm. But court record access policies that transgress neutral principles, especially with rules that are case non-specific, supra-political, and all-encompassing, call to mind the official abuses that prompted a FOIA revolution in the 1960s. The pendulum of transparency in government at that time swung out toward robust access. Developments in court record access and privacy might signal a pendulum swinging back.

NOTES

1. *Administrative Procedure Act of 1946*, Pub. L. No. 79-404, 60 Stat. 237 (1946).
2. *Electronic Freedom of Information Act Amendments of 1996*, Pub. L. No. 104-231, 110 Stat. 3048 (1996).
3. Richard J. Peltz-Steele, Joi Leonard, and Amanda J. Andrews, 2006. "The Arkansas Proposal on Access to Court Records: Upgrading the Common Law with Electronic Freedom of Information Norms," *Arkansas Law Review* 59 (555, 2006): 555–736.
4. John Losinger, "Electronic Access to Court Records: Shifting the Privacy Burden Away from Witnesses and Victims," *University of Baltimore Law Review* 26 (spring 2007): 419–49.
5. *Nixon v. Warner Communications, Inc.*, 435 U.S. 589 (1978).
6. Peltz-Steele, Leonard, and Andrews, "The Arkansas Proposal on Access to Court Records."
7. Charles N. Davis, "Expanding Privacy Rationales Under the Federal Freedom of Information Act: Stigmatization as Talisman," *Social Science Computer Review* 23 (winter 2005): 453–62; Daniel J. Solove, *The Digital Person* (New York, NY: New York University Press, 2004).
8. Peltz-Steele, Leonard, and Andrews, "The Arkansas Proposal on Access to Court Records."
9. Peter W. Martin, "Online Access to Court Records – From Documents to Data, Particulars to Patterns," *Villanova Law Review* 53 (1, 2008): 855–87.
10. Timothy B. Lee, 2009. "The Case Against PACER: Tearing Down the Courts' Paywall," *Ars Technica*, April 2009. http://www.arstechnica.com/tech-policy/news/2009/04/case-against-pacer.ars; Bruce Moyer, "Is PACER in Need of an Overhaul?" *Federal Lawyer*, August 2009, 10. N.J. Ct. Order 11–0017; Susan M. Thurston, "Pilot Program Expanded for Online Access to Court Digital Audio Recordings," *American Bankruptcy Institute Journal* 28 (June, 2009): 40–41, 68.
11. Peter A. Winn, "Judicial Information Management in an Electronic Age: Old Standards, New Challenges," *Federal Courts Law Review* 3 (July 2009): 135–76.
12. U.S. Courts, "Development of the Judicial Conference Policy on Privacy and Public Access to Electronic Case Files," http://www.uscourts.gov/RulesAndPolicies/JudiciaryPrivacyPolicy/DevelopmentOfPrivacyPolicy.aspx
13. Martha Wade Steketee and Alan Carlson, "Developing CCJ/COSCA Guidelines for Public Access to Court Records: A National Project to Assist State Courts," *The Justice Management*

Institute, 2002, http://www.jmijustice.org/publications/ccj-cosca-guidelines-for-public-access-to-court-records/view

14. Charles N. Davis, "Reconciling Privacy and Access Interests in E-government," *International Journal of Public Administration* 28 (7–8, 2005): 567–80; Charles N. Davis, "Revenge of the Pajama Surfers: The Inevitable Clash of E-governance and Informational Privacy Over Online Court Records," in *Handbook of Public Information Systems*, ed. Christopher M. Shea and G. David Garson, (Boca Raton, FL: CRC Press, 2010), 89–101; Peltz-Steele, Leonard, and Andrews, "The Arkansas Proposal on Access to Court Records."
15. National Center for State Courts, "Privacy/Public Access to Court Records: State Links," http://www.ncsc.org/topics/access-and-fairness/privacy-public-access-to-court-records/state-links.aspx?cat=Privacy%20Policies%20for%20Court%20Records
16. John Dooley, "State Policy Updates," Panel presentation at the 8th Conference on Privacy and Public Access to Court Records, Williamsburg, VA, November 3, 2011; R. Green, "Privacy and Open Courts: An Academic Perspective." Panel presentation at the 8th Conference on Privacy and Public Access to Court Records, Williamsburg, VA, November 3, 2011.
17. Commission on Public Access to Court Records, *Report to the Chief Judge of the State of New York*. February 25, 2004; http://www.nycourts.gov/ip/publicaccess/Report_PublicAccess_Court-Records.pdf
18. *E-Government Act of 2002*, Pub. L. No. 107-347, 116 Stat. 2899 (2002).
19. Fed. R. App. P. 25, 2007; Fed. R. Bankr. P. 9037, 2007; Fed. R. Civ. P. 5.2, 2007; Fed. R. Crim. P. 49.1, 2007.
20. Margaret Dee McGarity, "Privacy and Litigation: Two Mutually Exclusive Concepts," *Journal of American Academy of Matrimonial Lawyers* 23 (1, 2010): 99–109.
21. Peltz-Steele, Leonard, and Andrews, "The Arkansas Proposal on Access to Court Records."
22. *U.S. v. Amodeo*, 71 F. 3d 1044 (2d Cir. 1995).
23. Seymour Moskowitz, "Discovering Discovery: Non-Party Access to Pretrial Information in the Federal Courts 1938–2006," *University of Colorado Law Review* 78 (summer 2007): 817–78.
24. Kate Welsh, "Privacy and Courts: An International Perspective," Panel presentation at 8th Conference on Privacy and Public Access to Court Records, Williamsburg, VA (Nov. 3–4, 2011). *Whalen v. Roe*, 429 U.S. 589 (1977).
25. Steketee and Carlson, "Developing CCJ/COSCA Guidelines for Public Access to Court Records."
26. Peltz-Steele, Leonard, and Andrews, "The Arkansas Proposal on Access to Court Records."
27. James M. Chadwick, *Access to electronic court records: An outline of issues and legal analysis*, U.S. Department of Justice, http://it.ojp.gov/docdownloader.aspx?ddid=1024 (n.d.); Kristin M. Makar, "Taming Technology in the Context of the Public Access Doctrine: New Jersey Amended Rule 1:3," *Seton Hall Law Review* 41 (3, 2011): 1071–1109.
28. Winn, "Judicial Information Management in an Electronic Age: Old Standards, New Challenges."
29. Caren Myers Morrison, "Privacy, Accountability, and the Cooperating Defendant: Towards a New Role for Internet Access to Court Records," *Vanderbilt Law Review* 62 (April 2009): 921–78; David L. Snyder, Nonparty Remote Electronic Access to Plea Agreements in the Second Circuit. *Fordham Urban Law Journal* 35 (October 2008): 1263–1307.
30. Peltz-Steele, Leonard, and Andrews, "The Arkansas Proposal on Access to Court Records."
31. Leah Gurowitz, "Family Court and the Press: Privacy and Court Records." Panel presentation at 8th Conference on Privacy and Public Access to Court Records, Williamsburg, VA, Nobember 4, 2011.

32. Larry Smukler, "Family Court and the Press: Privacy and Court Records," Panel presentation at 8th Conference on Privacy and Public Access to Court Records, Williamsburg, VA, November 4, 2011.
33. Rebecca Hulse, "Privacy and Domestic Violence in Court," *William and Mary Journal of Women and the Law* 16 (2, 2010): 237–89; Helen Nissenbaum, "Privacy and Open Courts: An Academic Perspective," Panel presentation at 8th Conference on Privacy and Public Access to Court Records, Williamsburg, VA, November 3, 2011.
34. Privacy and Open Courts: An Academic Perspective, Audience comment at panel presentation at 8th Conference on Privacy and Public Access to Court Records, Williamsburg, VA, November 3, 2011.
35. Alan Carlson, *Technology issues and trends*. Panel presentation at the 8th Conference on Privacy and Public Access to Court Records, Williamsburg, VA, November 3, 2011.
36. Losinger, "Electronic Access to Court Records: Shifting the Privacy Burden Away from Witnesses and Victims."
37. Wendell Skidgel, "Technology Issues and Trends," Panel presentation at 8th Conference on Privacy and Public Access to Court Records, Williamsburg, VA, November 3, 2011.
38. U.S. Courts, "Judicial Conference Policy on Privacy and Public Access to Electronic Case Files," (March 2008). http://www.privacy.uscourts.gov/privacypolicy_Mar2008Revised.htm
39. Green, "Privacy and Open Courts: An Academic Perspective"; *In Re: Privacy and Public Access to Court Documents*, Case No. SCAD-2008-23, 2008 Okla. LEXIS 24 (March 11, 2008).
40. Mo. Ct. Order 11-0016, 2011.
41. Joseph F. Anderson, "Secrecy in the Courts: At the Tipping Point?" *Villanova Law Review* 53 (5, 2008): 811–28; Dooley, "State Policy Updates."
42. P. O'Donnell, "State Policy Updates," Panel presentation at 8th Conference on Privacy and Public Access to Court Records, Williamsburg, VA, November 3, 2011; Cal. R. Ct. 2.503. 2012.
43. Dooley, "State Policy Updates."
44. Cal. R. Ct. 2.503(e), (i) 2012.
45. cf. Fed. R. Civ. Pro. 5.2.
46. Michael Johnson, "Evolving Issues: Social Media and the Courts." Panel presentation at 8th Conference on Privacy and Public Access to Court Records, Williamsburg, VA, November 3, 2011; Minn. Juv. Prot. P. R. 8.06, 2012.
47. Makar, "Taming Technology in the Context of the Public Access Doctrine"; J. Perez, "State Policy Updates," Panel presentation at 8th Conference on Privacy and Public Access to Court Records, Williamsburg, VA, November 3, 2011.
48. Peltz-Steele, Leonard, and Andrews, "The Arkansas Proposal on Access to Court Records."
49. *U.S. Department of Justice v. Reporters Committee for Freedom of the Press*, 489 U.S. 749 (1989): 762.
50. Charles N. Davis, "Electronic Access to Information and the Privacy Paradox: Rethinking Practical Obscurity and its Impact on Electronic Freedom of Information," *Social Science Computer Review* 21 (February 2003): 15–25.
51. Helen Nissenbaum, *Privacy in Context: Technology, Policy, and the Integrity of Social Life* (Stanford, CA: Stanford Law Books, 2010).
52. McGarity, "Privacy and Litigation: Two Mutually Exclusive Concepts."
53. *Los Angeles Police Department v. United Reporting Publishing Corp.*, 528 U.S. 32 (1999).
54. Peltz-Steele, Leonard, and Andrews, "The Arkansas Proposal on Access to Court Records."
55. John J. Watkins and Rihard J. Peltz, *The Arkansas Freedom of Information Act*. 5th ed. (Fayetteville, AR: Arkansas Law Press, 2010).

56. Judith Resnik, "Courts: In and Out of Sight, Site, and Cite," *Villanova Law Review* 53 (2008): 771–810.
57. Lucy Dalglish, "General Discussion on Privacy and Public Access to Court Files," *Fordham Law Review* 79 (1, 2011): 1–24; Robert Gellman, "Courts Face Guilty Verdict on Bad Privacy, Public Access Policies," *Government Computer News*, February 4, 2002, 25; Lynn M. LoPucki, "Court-System Transparency," *Iowa Law Review* 94 (February 2009): 481–538.
58. Moyer, "Is PACER in Need of an Overhaul?"
59. J. James Christian, "Federal-Court Records via PACER and RECAP," *Arizona Attorney* 46 (November 2009): 24–26; RECAP Firefox Extension Home Page, http://www.recapthelaw.org.
60. Public Access to Court Electronic Records, "Acknowledgment of Policies and Procedures," May 17, 2010. http://www.pacer.gov/documents/pacer_policy.pdf.
61. Steketee and Carlson, "Developing CCJ/COSCA Guidelines for Public Access to Court Records."
62. Peltz-Steele, Leonard, and Andrews, "The Arkansas Proposal on Access to Court Records."
63. Cal. R. Ct. 2.503(f)–(g) 2012.
64. George F. Carpinello, *State policy updates*. Panel presentation at 8th Conference on Privacy and Public Access to Court Records, Williamsburg, VA, November 3, 2011.
65. Ark. S. Ct. Admin. Order 19. 2008, https://courts.arkansas.gov/rules/admin_orders_sc/admord19.pdf; Peltz-Steele, Leonard, and Andrews, "The Arkansas Proposal on Access to Court Records."
66. Peltz-Steele, Leonard, and Andrews, "The Arkansas Proposal on Access to Court Records."
67. Moskowitz, "Discovering Discovery: Non-Party Access to Pretrial Information in the Federal Courts 1938–2006."
68. *Griswold v. Connecticut*, 381 U.S. 479 (1965).
69. *Whalen v. Roe*, 429 U.S. 589 (1977).
70. Ross E. Cheit, "Tort Litigation, Transparency, and the Public Interest," *Roger Williams University Law Review* 13 (winter 2008): 232–84; T.S. Ellis, "Systematic Justice: Sealing, Judicial Transparency and Judicial Independence," *Villanova Law Review* 53 (5, 2008): 939–50.
71. Cheit, "Tort Litigation, Transparency, and the Public Interest"; Winn, "Judicial Information Management in an Electronic Age."
72. U.S. Courts, "Judicial Conference Policy on Sealed Cases," (2011) http://www.uscourts.gov/uscourts/News/2011/docs/JudicialConferencePolicyOnSealedCivilCases2011.pdf; Laurie Dore, Ronald J. Hedges, and Kenneth J. Withers (Eds.), "The Sedona Guidelines: Best Practices Addressing Protective Orders, Confidentiality and Public Access in Civil Cases," *Sedona Conference Journal* 8 (September 2007): 141–88.
73. Peltz-Steele, Leonard, and Andrews, "The Arkansas Proposal on Access to Court Records."
74. *Press-Enterprise Co. v. Superior Court*, 478 U.S. 1 (1986).
75. Peltz-Steele, Leonard, and Andrews, "The Arkansas Proposal on Access to Court Records"; Robert T. Reagan, "Sealing Court Records and Proceedings: A Pocket Guide," http://cryptome.org/0003/sealing-guide.pdf.
76. *Associated Press v. New Hampshire*, 888 A. 2d 1236 (N.H. 2005).
77. Peltz-Steele, Leonard, and Andrews, "The Arkansas Proposal on Access to Court Records."
78. N.J. Ct. Order 11-0017 2011.
79. e.g. Fed. R. Civ. P. 5.2.
80. Dalglish, "General Discussion on Privacy and Public Access to Court Files."
81. Peltz-Steele, Leonard, and Andrews, "The Arkansas Proposal on Access to Court Records."
82. Karen Eltis, "The Judicial System in the Digital Age: Revisiting the Relationship Between Privacy and Accessibility in the Cyber Context," *McGill Law Journal* 56 (February 2011): 289–316.

83. Dore, Hedges, and Withers, "The Sedona Guidelines: Best Practices Addressing Protective Orders, Confidentiality and Public Access in Civil Cases"; Winn, "Judicial Information Management in an Electronic Age: Old Standards, New Challenges."
84. N.J. Ct. Order 11–0017 2011.
85. Resnik, "Courts: In and Out of Sight, Site, and Cite."
86. Janet Walker and Garry D. Watson, "New Trends in Procedural Law: New Technologies and the Civil Litigation Process," *Hastings International and Comparative Law Review* 31 (October 2008): 251–93.
87. Anderson, "Secrecy in the Courts: At the Tipping Point?"
88. Welsh, "Privacy and Courts: An International Perspective."
89. Peter S. Kozinets, "Access to Metadata in Public Records: Ensuring Open Government in the Information Age," *Communications Lawyer* 27 (July 2010): 1, 23–29; *Arnold v. Pennsylvania Department of Transportation*, 477 F. 3d 105 (3d Cir. 2007); *Pintos v. Pacific Creditors Association*, 605 F. 3d 665 (9th Cir. 2010).
90. Daniel J. Solove, "Access and Aggregation: Public Records, Privacy and the Constitution," *Minnesota Law Review* 86 (6, 2002): 1137–1218.
91. Meliah Thomas, "The First Amendment Right of Access to Docket Sheets," *California Law Review* 94 (5, 2006): 1537–80.
92. e.g. *Lugosch v. Pyramid Co.*, 435 F. 3d 110 (2d Cir. 2006).
93. Peltz-Steele, Leonard, and Andrews, "The Arkansas Proposal on Access to Court Records."
94. Harold L. Cross, *The People's Right to Know: Legal Access to Public Records and Proceedings*. (Morningside Heights, NY: Columbia University Press, 1953): 49 & app. 3, 15.
95. Cross, *The People's Right to Know*, 16.
96. Cross, *The People's Right to Know*, 8.
97. U.S. Congress. House. 89th Cong., 2d Sess., 1966. H.R. Rep. No. 1497.
98. *Freedom of Information Act of 1966*, 5 U.S.C. §552 (2009): 23.
99. *Freedom of Information Act*, 30.
100. U.S. Congress. House. 104th Cong., 2d Sess., 1996. H.R. Rep. No. 104-795; U.S. Congress. Senate. 104th Cong., 1st Sess., 1995 S. Rep. No. 104-272.
101. Peltz-Steele, Leonard, and Andrews, "The Arkansas Proposal on Access to Court Records."
102. Gregg Leslie and Mark Caramanica (eds.). *Open Government Guide: Access to Public Records and Meetings* (6th ed., 2011). http://www.rcfp.org/open-government-guide
103. e.g., Conn. Gen. Stat. §1-210(b)(18) 2011.
104. Leslie and Caramanica, *Open Government Guide*.
105. e.g., Ga. Code Ann. §50-18-72(4.1) 2010.
106. e.g., S.D. Code §1-27-1.11 2011.
107. Leslie and Caramanica, *Open Government Guide*.
108. Kristen M. Blankley, "Are Public Records too Public? Why Personally Identifying Information Should be Removed from Both Online and Print Versions of Court Documents," *Ohio State Law Journal* 65 (3, 2004): 413–50.
109. Charles N. Davis, "Expanding Privacy Rationales Under the Federal Freedom of Information Act: Stigmatization as Talisman," In Todd Loendorf and G. David Garson (Eds.), *Patriotic information systems* (Hershey, Pa.: IGI Global, 2008): 42–56; Watkins and Peltz, *The Arkansas Freedom of Information Act*.
110. Daniel J. Solove, *Understanding Privacy* (Cambridge, MA: Harvard University Press, 2008): 111.

111. Damon Greer, "Privacy in the Post-Modern Era—An Unrealized Ideal? *Sedona Conference Journal* 12 (September 2011): 189–202.
112. Solove, *Understanding Privacy.*
113. Nissenbaum, "Privacy and Open Courts."
114. Solove, *Understanding Privacy*, 117, 129, 149.
115. Nissenbaum, "Privacy and Open Courts," 148.
116. Nissenbaum, "Privacy and Open Courts," 221.
117. *FAA v. Cooper*, 131 S. Ct. 3025 (2011).
118. *U.S. v. Jones*, Case No. 10-1259, 2012 WL 171117 (U.S. Jan. 23, 2012).
119. *U.S. v. Jones*, 10.
120. *U.S. v. Custer Battlefield Museum & Store Located at Interstate 90, Exit 514, South of Billings, Mont.*, 658 F. 3d 1188 (9th Cir. 2011).
121. Colo. Rev. Stat. Ann. §18-3-407, 2006.
122. Lawrence Lessig, *Code and Other Laws.* (New York, NY: Basic Books, 1999).

CHAPTER SIX

Social Media and Reporting on Judicial Proceedings: A Digital Era Conflict

DERIGAN SILVER

In February 2009, U.S. District Judge J. Thomas Marten became the first federal judge to officially allow a journalist to use a Blackberry device to update his Twitter account from within a federal courtroom when he granted Ron Sylvester permission to use the microblogging site to cover a federal racketeering trial.[1] While Sylvester, then a reporter for the Wichita Eagle in Wichita, Kansas, had been tweeting from the courtroom since his 2007 coverage of a murder case, Marten's decision was the first official order from a federal judge allowing this form of coverage. Not long after, in 2010 Chief Judge Vaughn R. Walker of the Northern District of California allowed reporters to tweet from the courtroom during a lawsuit involving a challenge to California's Proposition 8, which amended the California Constitution to include a section providing that the state would only recognize marriage between a man and a woman.[2]

While some commentators found these decisions less than shocking—after all, as long ago as the Scooter Libby Trial the U.S. District Court in Washington had given the Media Bloggers Association two press credentials—not long after Marten's order was announced, U.S. District Judge Clay Land ruled that Rule 53 of the Federal Rules of Criminal Procedure should be interpreted as banning tweeting. According to Land, the rule's prohibition on the "broadcasting of judicial proceedings" included "sending electronic messages from a courtroom that contemporaneously describe the trial proceedings and are instantaneously available for public viewing."[3] Thus, the three judges' conflicting decisions indicate the discussion over social media and access to judicial proceedings is just beginning, a discussion made more urgent by the rapid adoption of Twitter.

A study by the Pew Internet & American Life Project showed that one in five people ages 18–24 have used Twitter, and that people who use Twitter are more likely to read the news on a cell phone, smartphone, or web site than to read a traditional printed newspaper. In response, many media organizations are bolstering their traditional coverage with social media. For example, one author compared the *New York Times*' embrace of social media to "saturation bombing in cyberspace."[4] The use of Twitter to cover trials has increased as well. As one commentator recently wrote, "Tweeting from the courtroom is de rigueur" for courtroom reporters.[5] Thus, as journalists' use of social media continues to grow, state and federal rules governing broadcasting from courtrooms will have to be interpreted and/or modified to account for new technologies.

The purpose of this chapter is to outline how courts are changing their rules to deal with social media and review the intersection of social media and government transparency laws. It is important to note that it focuses on rules governing non-participants; different, often much stricter, rules apply to trial participants such as lawyers, court officials, jurors, and witnesses. The chapter begins by reviewing Supreme Court cases that have established a First Amendment right to attend trials. Next it examines rules regarding cameras and social media in the courtroom, first explaining the rules that govern federal courtrooms and then those that govern the state court system. Although the chapter attempts to provide guidance, unfortunately for journalists there appears to be no set standard regarding tweeting from courtrooms, and the rules tend to vary, at times, from one trial to the next.

ACCESS TO COURTS

The U.S. Supreme Court has found that the First Amendment guarantees a broad right of access to judicial proceedings and documents. In 1980 the Court began limiting judges' ability to close access to their courtrooms by ruling in *Richmond Newspapers v. Virginia*[6] that the press and public have a First Amendment right to attend criminal trials. In 1978, after Circuit Court Judge Richard Taylor ordered the murder trial of John Paul Stevenson closed, Richmond Newspapers protested and asked for a hearing, which Judge Taylor conducted in secret. At the conclusion of the hearing, Taylor stuck by his decision to close the trial. The next day Judge Taylor dismissed the jury, ruled the state had not presented sufficient evidence to justify a guilty verdict and declared Stevenson not guilty. The decision was reported in a two-sentence order, and, because the trial had been conducted behind closed doors, there was no way for the public to determine if Judge Taylor's conclusion was justified. Richmond Newspapers appealed the trial closure to the

Virginia Supreme Court, which refused to overturn Judge Taylor's decision. The U.S. Supreme Court, however, ruled 7-1 that the closure was unconstitutional.

In an opinion by Chief Justice Burger, the Court held the public right of access to a criminal trial could be overcome only by "an overriding interest," and only if the trial judge finds that "alternative solutions" are inadequate to ensure fairness. In a concurring opinion, which later became very influential in the development of a test for determining when courtrooms could be closed, Justice William Brennan stressed the structural value of open courts, that is, the role openness plays in self-government.

Two years after *Richmond Newspapers*, the Court ruled 6-3 that a blanket rule requiring courtroom closure during the testimony of minor victims in sex crime cases violated the First Amendment. A trial judge in Norfolk County, Massachusetts, had closed the trial of a man charged with raping two 16-year-olds and one 17-year-old girl. The judge ruled a state statute mandated closing the entire trial. In *Globe Newspaper Co. v. Superior Court*,[7] Justice Brennan, writing for the majority, concluded the closure of criminal trials had to be determined on a case-by-case basis. While Justice Brennan conceded that protecting "minor victims of sex crimes from further trauma and embarrassment" was a compelling interest, he wrote it did not justify a mandatory closure law since such a measure was not narrowly tailored.

The Court expanded access to other courtroom proceedings in the 1980s. In 1984, the Court ruled that as an integral part of a criminal trial, jury selection is subject to the First Amendment presumption of access. *Press-Enterprise Co. v. Riverside County Superior Court*[8]—now known as Press-Enterprise I to distinguish it from a later case of the same name—began when a California judge closed almost six weeks of voir dire (jury selection proceedings) in a rape and murder trial involving a teenage victim. After the defendant was convicted and sentenced to death, the judge still refused to release the transcript of the voir dire. The Supreme Court unanimously ruled that closure of the voir dire violated the First Amendment. In 1986, the Supreme Court extended the First Amendment presumption of openness to a criminal pretrial proceeding, a preliminary hearing. In some states, a preliminary hearing is known as a "show-cause" hearing and is designed to determine if the government has sufficient evidence to bind a defendant over for trial. *Press-Enterprise v. Riverside County Superior Court*,[9] or *Press-Enterprise II*, resulted from the closure of a 41-day preliminary hearing in the case of Robert Diaz, a nurse accused of murdering twelve patients by administering massive doses of a heart drug.

Various courts have also held that a constitutional right of access attaches to numerous other types of criminal proceedings, including plea and sentencing hearings, bail hearings, and a post-trial examination of jurors about potential misconduct. However, some types of criminal proceedings, most notably grand jury proceedings, have historically been conducted behind closed doors, and courts continue to rule that the First Amendment presumption of openness does not apply. In addition, juvenile proceedings have historically been closed to the public. State laws often provided for closed hearings and confidential records in juvenile proceedings to avoid stigmatizing minors by making their identities and crimes known to the public. Prompted by a rash of high profile, serious crimes by juveniles during the 1990s, however, many states began opening their juvenile proceedings and records to the public and press. Several lower courts have held that the First Amendment right of access does not apply to such proceedings. On the other hand, other courts have found a constitutional right of public and press access to juvenile proceedings—either based on the First Amendment or on state constitutional provisions. The Supreme Court has yet to rule whether the First Amendment right applies to juvenile proceedings,

Similarly, although the Supreme Court has never ruled that civil trials are subject to the same First Amendment right of access as criminal proceedings, lower courts have held that the press and public enjoy a right of access to civil judicial proceedings. Some courts have based this right of access on the First Amendment. Other courts have found a common law or state constitutional basis for access. Not every type of civil proceeding is open to the press and public in every jurisdiction, however. For example, involuntary commitment and adoption proceedings are frequently cited as two of the most commonly closed civil proceedings. Furthermore, just as with criminal proceedings, judges sometimes find that even when a right of access exists for a particular type of civil proceeding, a compelling governmental interest can overcome that.

Civil discovery proceedings have generated a number of access lawsuits. Discovery, the pretrial phase of a lawsuit during which the litigants attempt to collect the information they need from one another and from third-party witnesses, has not been open to the press and public.

USING TWITTER IN FEDERAL COURTS

While the Supreme Court and lower courts have ruled the press and the public are allowed to attend trials, the question of whether courtroom proceedings can be broadcast is a different topic, one which has been taken up by both courts and legislatures. The ability to broadcast from court depends both on the type of court and the type of case being heard. In 1976 only two states—Texas and Colorado—allowed cameras in their courtrooms. By 2010 every state allowed at least some camera coverage. A key reason for this shift is technological advances. A camera is no longer a bulky apparatus with obtrusive lights and electrical cords snaking across courtroom floors. Today, a camera can literally fit in one's pocket. Changing public and judicial attitudes about cameras have also been important. Over the last thirty years, video cameras have become part of everyday life. Most people are accustomed to being videotaped and many of us have posted videos of ourselves, our family or our friends on Youtube or Facebook.

The situation was significantly different when cameras were obvious and intrusive. The first official bans on cameras in the courtroom followed the 1935 trial of Bruno Hauptmann for the kidnapping of the Lindbergh baby. Although the New Jersey appellate court found no error in the actions of the media during the trial in 1937, in response to the case the American Bar Association adopted Canon 35 as part of its Canons of Professional and Judicial Ethics. Canon 35, which declared that cameras "detract from the essential dignity of the proceedings, degrade the court and create misconceptions with respect thereto in the mind of the public," recommended that cameras be banned in all courtrooms. That recommendation was amended in 1952 to specifically include television cameras. When the American Bar Association replaced the Canons of Professional and Judicial Ethics with the Code of Judicial Conduct in 1972 the ban on television cameras continued. In 1946, the Federal Rules of Criminal Procedure adopted a rule prohibiting the taking of photographs and the radio broadcasting of all criminal proceedings. Rule 53 of the Federal Rules of Criminal Procedure currently holds that "except as otherwise provided by a statute or these rules, the court must not permit the taking of photographs in the courtroom during judicial proceedings or the broadcasting of judicial proceedings from the courtroom."

In 1965, when the Supreme Court heard its first cameras-in-the-courtroom case, *Estes v. Texas*,[10] it agreed with the ABA that television cameras intruded upon "the solemn decorum of court procedure." The Court overturned a criminal conviction under the Due Process Clause when the proceedings were broadcast over the petitioner's objection. Justice Tom Clark held that jurors could be distracted by the equipment and by the mere knowledge that the trial was being televised. The judge in the case, Clark wrote, had enough to do supervising the trial without also worrying about supervising television crews. Justice John Marshall Harlan wrote a concurring opinion in which he explained his fears about television's potential for causing "serious mischief." Harlan, however, also speculated that someday television might become so commonplace its disruptive influence would disappear. Justice Harlan wrote "that the day may come when television broadcasting will have become so commonplace an affair in the daily life of the average person as to dissipate all reasonable likelihood that its use in the courtrooms may disparage the judicial process."

In its second cameras-in-the-courtroom case, *Chandler v. Florida*,[11] the Supreme Court held that "no one has been able to present empirical data sufficient to establish that the mere presence of the broadcast media inherently has an adverse effect on that process." Florida was one of several states that had begun experimenting with allowing cameras in its courtrooms in the 1970s. In 1978, the American Bar Association's Committee on Trial-Free Press proposed allowing broadcasting when coverage would not be obtrusive. In response, Florida adopted a canon which permitted electronic media and still photography. In *Chandler*, which involved two Miami Beach police officers charged with burglarizing a restaurant, the judge allowed a television camera in the court despite the defendants' objections. Although only about three minutes of the trial were ultimately broadcast, the officers appealed their conviction.

The Court, however, ruled the mere presence of cameras in the courtroom did not automatically deprive a defendant of a fair trial. Chief Justice Warren Burger wrote that broadcast coverage might, in some cases, violate a defendant's constitutional rights, but the mere possibility of prejudice did not justify a ban on televising trials. Instead, individual defendants had to prove that the presence of cameras impaired the ability of the jurors to decide the specific case. It's important to note, however, that while *Chandler v. Florida* permits states to open their courtrooms to cameras, it does not require them to do so. The Supreme Court did not say that photographers have a First Amendment right to take their equipment into courtrooms.

Today, federal courts generally prohibit all broadcasting, while many state courts often permit it. The Federal Rules of Criminal Procedure have prohibited televising of federal criminal trials for more than fifty years. In 1983, in *United States v. Hastings*,[12] the United States Court of Appeals for the Eleventh Circuit affirmed that Federal Rule 53 did not violate the First Amendment. In Hastings, several media organizations and the defendant in the case wanted the trial court to televise proceedings. Discussing *Globe Newspapers*, the court concluded the right to attend criminal trials could not be extended to the right to televise, record, or broadcast trials. In 1996, the Judicial Conference of the United States announced a prohibition on electronic media coverage of civil and criminal trials, while the Circuit Council adopted a policy of allowing federal appellate court judges to allow broadcasting at their discretion.

In 2011 the federal government announced a new three-year pilot program that placed cameras in 14 federal trials. The cameras are owned by the courts and controlled by judges. The proceedings will not be instantly available to the media and will instead only be available on the courts' website at a later time. Only civil proceedings will be recorded, and both parties and the judge must consent to the recording. In addition, the presiding judge may terminate recordings in the interest of justice, to protect the dignity of the court and the rights of the parties and witnesses, or for any reason the judge considers "necessary or appropriate."[13] Over the past few years Congress has repeatedly considered legislation to permit federal judges, both trial and appellate, to open their courtrooms to cameras. However, no camera access bill has managed to pass both houses.

The advent of new media technologies has not necessarily changed the federal judiciary's stance on cameras. In 2010, the Supreme Court barred a federal district court from broadcasting *Hollingsworth v. Perry*,[14] the non-jury federal trial in a case involving California's Proposition 8 discussed in the introduction to this chapter. The Ninth Circuit Judicial Council decided to begin a pilot program allowing for the limited use of cameras in federal district courts in the Circuit. On January 6, 2010, in addition to allowing reporters to live tweet the Proposition 8 trial, Judge Walker also announced that under the pilot program the trial would be streamed live to courthouses in other cities and recorded for broadcast on the Internet. Supporters of Proposition 8 filed an application for a stay of the decision with the Supreme Court. In a per curiam opinion the Court wrote that while it was not "expressing any views on the propriety of broadcasting court proceedings generally," the majority held that the district court had not properly amended its local rules to allow for broadcasting of the trial in accordance with federal law. The Ninth Circuit, in its gradual acceptance toward broadcasting trials, decided to voluntarily begin broadcasting live video streaming of important cases on its website, starting December 2013.[15]

While movement is afoot to allow broadcasting of trials, federal courts have been slow to embrace Twitter. As discussed above, in February 2009, Judge J. Thomas Marten allowed Ron Sylvester, a reporter for the *Wichita Eagle* in Wichita, Kansas, to cover the trial of six suspected gang members in snippets of 140 characters. While Sylvester had been tweeting from the courtroom since 2007, this was the first official order from a judge allowing this form of coverage. Some of Sylvester's tweets included:

- "Judge Marten is talking to reluctant witness in chambers with a court reporter transcribing the conversation."
- "The witness who was yelling in the hallway earlier has not returned to the courthouse."
- "Defendants are chatting and laughing among themselves."
- "Exhibits are shown electronically. Every juror has a monitor in the box. There is a monitor at each lawyer's table and one for the gallery."[16]

Most federal courts, however, have not addressed the issue of Twitter, although some have interpreted the ban on broadcasting in federal courts to now include the use of Twitter. In *United States v. Shelnutt*,[17] a reporter for the *Columbus Ledger-Enquirer* was denied a request to use a handheld electronic device to send tweets to his newspaper's Twitter page. In the case, the District Court for the Middle District of Georgia held that reporters could not tweet during trial because the activity violated Rule 53. Using Webster's Third New International Dictionary to define the word "broadcast," the court concluded the definition of broadcasting included "casting or scattering in all directions" and "the act of making widely known."[18] The court wrote, "It cannot be reasonably disputed that 'twittering' ... would result in casting to the general public and thus making widely known the trial proceedings."[19] The court thus concluded "broadcasting" includes the sending of electronic messages that contemporaneously describe the trial proceedings and are instantly available.

Similarly, Chief U.S. District Court Judge David C. Norton of South Carolina signed an order in April 2011 that banned all wireless communication devices in courtroom facilities.[20] Most recently, in January 2012, U.S. District Judge Sara Lioi set a number of rules before the public corruption trial of former Cuyahoga County Commissioner Jimmy Dimora. The *Cleveland Plain Dealer* had planned for its reporters to tweet from a court-supervised media room. Lioi, however, banned reporters from tweeting and live blogging from the courthouse during the trial when Dimora's lawyers objected to the practice.[21]

While these examples demonstrate some judges' reluctance to allow tweeting from courts, other judges have embraced it. For example, in response to a request to live tweet a tax fraud trial, U.S. District Court Judge Mark Bennett of the Northern District of Iowa simply asked a reporter from the Cedar Rapids *Gazette* to sit farther back in the courtroom so her typing would not be distracting.[22] In March 2009, U.S. District Judge Federico Moreno of the Southern District of Florida banned reporters from posting live web updates from the courtroom, but allowed them to step outside into the hallway to do so.[23]

Unfortunately, to date the U.S. Circuit Courts of Appeals have remained silent on the issue of tweeting or using wireless communication devices. Thus, before covering a proceeding in federal court reporters should consult with court personnel or know the presiding judge's individual preferences. Before tweeting from a judge's courtroom, check the court's standing orders and see if the court's electronic device policy is posted. A courthouse-wide standing order or electronic device policy should be located on the courthouse's website, while an individual judge's preferences might be located on a page dedicated to that judge. These materials might also contain information on how to get permission to use a device. If it's unclear how the court will apply rules designed to deal with traditional electronic media to live-blogging or what a judge's individual preference might be journalists should contact the court's Public Information Officer (PIO).

USING TWITTER IN STATE COURTS

While the federal courts have resisted efforts to bring cameras into courtrooms, almost all states allow some degree of camera coverage. Rules regulating broadcasting in courtroom vary state by state, however, and the extent of coverage permitted, as well as the specific rules journalists must follow, differs greatly among the states. The best source for information about cameras-in-the-courtroom rules in the 50 states is an online guide provided by the Radio Television Digital News Association.[24]

In addition, most states have yet to consider the question of whether updating a Twitter page is "broadcasting." Even states that have traditionally been camera friendly may have trouble adjusting to new media such as Twitter. For example, in Florida, home of the Casey Anthony trial, electronic media and still photography coverage of proceedings is allowed, subject only to the authority of the presiding judge. In early 2010, however, the Standing Committee on Rules of Practices and Procedures considered a statewide ban on electronic devices in courthouses. The proposal was voted down by the committee 11 to 5.[25]

Thus, as is true with federal courts, it is important journalists know their individual state rules and a judge's individual preferences before taking a camera or a wireless communication device into a courtroom or using a smartphone to update a microblogging site such as Twitter. As in federal court, before covering the proceedings, journalists should check the court's standing orders, free-standing electronic device policies, and the presiding judge's individual preferences. Journalists should also keep in mind anyone who violates the rules governing electronic coverage can be found in contempt of court. It is also important to remember that in almost every state that allows broadcasting from proceedings, permission must be obtained from the presiding judge in advance.

To date, two states, Utah and Kansas, have officially adopted statewide rules that allow journalists to live stream via laptops and phones from state courtrooms with the presiding judge's permission. In October 2012, the Kansas Supreme Court amended its rules to clarify that journalists, with the permission of the presiding judge, may use laptops and cell phones to tweet or update other microblogging sites. Under the rule, journalists are still prohibited from taking pictures of jurors, juveniles, and undercover agents.[26] Soon after, the Utah Supreme Court approved a rule that allows journalists to tweet, live stream, and blog from the courtroom. Under the new rule, electronic media coverage is permitted in proceedings that are open to the public, although journalists have to submit an application to the court before using an electronic device in a courtroom, and taking picture of jurors, minors or attorney documents is still prohibited.[27]

Nineteen states allow for a wide range of coverage, giving a great deal of discretion to the presiding judge to determine when to allow broadcasting. Typically, these states require requests that to broadcast from the trial be approved beforehand. In California, for example, where broadcasting is governed by Rule 1.150 of the California Rules of Court, media coverage is permitted by written order of the judge. A request for coverage must be filed at least five court days before the proceeding to be covered. Rule 1.150 does not create a presumption for or against granting permission, however, and leaves the use of cameras in the courtroom to the judge's discretion.[28] It is also important to note that even in these states taking pictures of members of the jury, witnesses, or juveniles might be prohibited.

States allowing broadcast at the judge's discretion

California, Colorado, Florida, Georgia, Idaho, Kentucky, Michigan, Montana, Nevada, New Hampshire, New Mexico, North Dakota, South Dakota, South Carolina, Tennessee, Vermont, Washington, West Virginia, Wisconsin, and Wyoming

Sixteen states prohibit coverage of specific cases or when witnesses object. For example, in Alaska, Administrative Rule 50 permits members of the electronic media to cover court proceedings in all state trial and appellate courts and is not limited to courtrooms. However, consent of the presiding judge is needed to cover a proceeding and the consent of all parties is required for coverage of divorce, dissolution of marriage, domestic violence, child custody and visitation, paternity, or other family proceedings. In Missouri, media coverage at both the trial and appellate levels is permitted, but broadcasting from jury selection, juvenile, adoption, domestic relations, and child custody cases is not, although the media may record a juvenile being prosecuted as an adult. Additionally, in many of these states jurors may not be photographed, filed or videotaped in the courtroom at any time.

North Carolina has one of the most interesting policies regarding broadcasting from courtrooms. According to Rule 15 of the General Rules of Practice for Superior and District Courts, "electronic media coverage" and "electronic coverage" is allowed, but if the location of equipment and personnel necessary for electronic media coverage is to be located within the courtroom, the area must be set apart by a booth or other partitioning device constructed therein at the expense of the media. North Carolina has yet to determine how this requirement might apply to a member of the media seeking to use an electronic device to update a Twitter page. Additionally, coverage is not allowed in adoption proceedings, juvenile proceedings, probable cause proceedings, child custody proceedings, divorce proceedings, temporary and permanent alimony proceedings, proceedings for the hearing of motions to suppress evidence, and proceedings involving trade secrets. Coverage of jurors is also expressly prohibited at any stage of a judicial proceeding.

States that prohibit coverage based on the type of case or when a witness objects

Alaska, Arizona, Connecticut, Hawaii, Indiana, Iowa, Kansas, Massachusetts, Missouri, North Carolina, New Jersey, Ohio, Oregon, Rhode Island, Texas, and Virginia

Fifteen states allow electronic coverage in appellate courts only or require the consent of all parties. New York has a ban on all televising of trial court proceedings, no matter what the circumstances or the assessment of the presiding judge. Some courts have recently begun pilot programs involving electronic media. For example, effective July 1, 2011, Minnesota began a two-year pilot project allowing a judge to authorize recording of court proceedings with the consent of all parties in criminal proceedings and without the consent of all parties in civil proceedings. At the appellate level, all-party consent is not required, but the Clerk of the Appellate Courts must be notified of intent to cover the proceedings at least 24 hours in advance.

States that allow appellate coverage only or require the consent of all parties

Alabama, Arkansas, Delaware, Illinois, Louisiana, Maine, Maryland, Minnesota, Mississippi, Nebraska, New York, Oklahoma, Pennsylvania, South Dakota, and Utah

While most states have yet to address the issue of live blogging from a courtroom, at least one state has recently moved to ban all cell phones from courtrooms, even though traditional broadcast media are allowed. In Maryland, electronic media coverage is allowed at civil trials upon written consent of all the parties. However, rule 16-110 requires cell phones and other electronic devices be turned off in all courtrooms and prohibits their use for video or photographs. Obviously, such a ban would include live blogging and tweeting.

CONCLUSION

As one observer noted, "(s)ocial networking websites have transformed the way people communicate and stay in touch with one another."[29] Coverage of courtroom proceedings is just one example of how new media can clash with old rules and regulations. The popularity of real-time communication technologies such as Twitter have introduced a new dimension into a journalist's coverage of court proceedings and the use of these technologies has been met with a mixture of both acceptance and criticism. The resultant changes in the methods and speed of communication promise to revolutionize the dissemination of information in ways that may change the very nature of what we mean by a "public trial." Journalists wishing to use these new technologies to reach their readers are advised to proceed with caution and never assume their use will be welcomed by everyone. While many judges have determined that Twitter is an effective tool that can bring even greater transparency to the judicial process, others have rejected the medium. A journalist should begin by determining what kind of proceeding they are interested in covering and the jurisdiction the proceeding will take place in. If the rules of that jurisdiction regarding Twitter or portable electronic devices are unclear, journalists should simply ask for permission to use the site. While one can hope that soon more judges will adopt the stance taken by U.S. District Judge Tom Marten that there is little difference between tweeting "and a journalist sitting there taking notes,"[30] the judicial system has not yet embraced Twitter in nearly the same way journalists have.

NOTES

1. Ahnalese Rushmann, "Courtroom Coverage in 140 Characters," *The News Media and the Law* 33 (Spring 2009): 28.
2. Joe Eskenazi, "Too Impatient for YouTube? Follow Prop 8 Trial on Twitter," *SF Weekly*, January 11, 2010.
3. *United States v. Shelnutt*, 2009 U.S. Dist. LEXIS 101427 (M.D. Ga. 2009).
4. Arielle Emmett, 2009. "Networking News: Traditional News Outlets Turn to Social Networking Web Sites in an Effort to Build Their Online Audiences," *American Journalism Review* 30 (December–January 2009): 40–43.
5. Nicole Lozare, "With 140 Characters at a Time, Twitter is Presenting New Challenges to Journalists," *The News Media and the Law* 35 (fall 2011): 4.
6. *Richmond Newspapers v. Virginia*, 448 U.S. 555 (1980).
7. *Globe Newspaper Co. v. Superior Court*, 457 U.S. 596 (1982).
8. *Press-Enterprise Co. v. Superior Court*, 464 U.S. 501 (1984).
9. *Press-Enterprise Co. v. Superior Court*, 478 U.S. 1 (1986).
10. *Estes v. Texas*, 381 U.S. 532 (1965).
11. *Chandler v. Florida*, 449 U.S. 560 (1981).
12. *United States v. Hastings*, 695 F.2d 1278 (11th Cir. 1983).
13. Judicial Conference Committee on Court Administration and Case Management Guidelines for The Cameras Pilot Project in the District Courts, 2011, http://www.uscourts.gov/uscourts/News/2011/docs/CamerasGuidelines.pdf
14. *Hollingsworth v. Perry*, 558 U.S. ___ (2010).
15. Court of Appeals to Open En Banc Proceedings to Internet Viewing, Ninth Circuit Court of Appeals, http://www.ce9.uscourts.gov/absolutenm/articlefiles/641-En_Banc_Streaming.pdf
16. Roxana Hegeman, "Twitter in the Court," *Editor and Publisher*, March 6, 2009.
17. *United States v. Shelnutt*.
18. *United States v. Shelnutt*, 2.
19. *United States v. Shelnutt*, 2.
20. Lozare, "With 140 Characters at a Time."
21. Rachel Dissell, "Federal Judge Bans Tweeting and Live Blogging from Jimmy Dimora Trial," *The Cleveland Plain Dealer*, January 3, 2012, http://blog.cleveland.com/metro/2012/01/federal_judge_bans_tweeting_an.html
22. Lozare, "With 140 Characters at a Time."
23. Rushmann, "Courtroom Coverage in 140 Characters."
24. Radio Television Digital News Association, "Cameras in the Court: A State-By-State Guide," http://www.rtnda.org/pages/media_items/cameras-in-the-court-a-state-by-state-guide55.php

25. Tricia Bishop, "New Rule Could End Tweets from Trials Statewide: Policy Would Bar Electronic Devices in Courthouses," *The Baltimore Sun*, February 22, 2010.
26. KS Sup. Ct. Rule 1001 Electronic and Photographic Coverage of Judicial Proceedings, available at http://www.kscourts.org/rules/Rule-Info.asp?r1=Media+Coverage+of+Judicial+Proceedings&-r2=318
27. Judicial Council Study Committee on Technology Brought into the Court, April 10, 2012, available at http://www.rcfp.org/sites/default/files/docs/20121120_133113_cameras_in_court.pdf
28. Adriana C. Cervantes, "Will Twitter be Following You in the Courtroom?: Why Reporters Should be Allowed to Broadcast During Courtroom Proceedings," *Hastings Communication and Entertainment Law Journal* 33 (fall 2010): 133–158.
29. Cervantes, "Will Twitter be Following You in the Courtroom?" 133.
30. Lynne Marek, "What is That Reporter Doing in Court? 'Twittering,'" *The National Law Journal*, March 16, 2009.

CHAPTER SEVEN

Access to Email and the Right of Privacy in the Workplace

KYU HO YOUM

The Internet is more pervasive than ever. Its impact on individuals and society is revolutionary. One of the most phenomenal developments in Internet communications is electronic mail or email. Email and its analogues are "the most used Internet facility."[1] More than 2.3 billion people are expected to use email, and 3.6 billion-plus email accounts were projected to be active worldwide in 2013.[2]

Email privacy is a major unsettled point of dispute over how to balance an individual's right of privacy with competing societal interests. One widely discussed example concerns email monitoring by employers in the workplace. Whether or not employees have a reasonable expectation of privacy in their work email, American courts have held that employers may monitor work emails when they have legitimate interests.[3]

Does or should email surveillance in the workplace extend to employees' private web-based email, which is separate from their work email? The underlying rationale for work email monitoring might apply to webmail monitoring. But the employees' privacy interests in webmail might not necessarily justify the comparable application of the case law.[4]

Public employees and private individuals can turn to the Fourth Amendment when the government monitors their emails. Private employees can invoke the federal Stored Communications Act (SCA). Unfortunately, the SCA and other statutes are often contradictory, confusing, and ambiguous. They also lag behind the "fast-changing" communication technology, and their updates have been "piecemeal."[5]

In the meantime, email privacy vs. government transparency has taken on a heightened urgency as a new issue for freedom of information. This is showcased when government officials' emails are requested under federal and state open records laws. Email poses an "additional challenge" to access-to-records compliance because government officials expect privacy in their email messages.[6]

Further, public officials' growing use of private email accounts for their official work has been a cause of concern for journalists and others.[7] In September 2012, for example, a think tank in Washington, D.C., sued the federal Environmental Protection Agency over access to agency officials' private emails.[8]

PRIVACY AS 'THE RIGHT TO BE LET ALONE'

As a malleable concept, privacy varies from individual to individual, from culture to culture, and from period to period. The United States often offers "a frame of reference" for international and foreign courts in balancing privacy with freedom of speech and the press.[9]

The word "privacy" does not appear in the Constitution of the United States. Yale law professor Akhil Amar, however, observes: "(W)hen we read between the lines and heed the document as a whole, with particular attention to its arc across the centuries, a different picture emerges."[10] Privacy is now entrenched as a right under the federal Constitution and the state constitutions. Further, it is protected by a number of federal and state statutes.

The U.S. Supreme Court has read privacy into various provisions of the federal Constitution. The Fourth Amendment provides for a "right of the people to be secure in their persons, houses, papers, and effects, against unreasonable searches and seizures."

In *Katz v. United States*,[11] the Supreme Court held: "What a person knowingly exposes to the public, even in his own home or office, is not a subject of Fourth Amendment protection. But what he seeks to preserve as private, even in an area accessible to the public, may be constitutionally protected."[12] The *Katz* case involved the FBI wiretapping a public phone booth. Most significantly, Justice John Harlan wrote in his influential concurring opinion that the Fourth Amendment should apply whenever a person exhibits an "actual (subjective) expectation of privacy" that "society is prepared to recognize as 'reasonable.'"[13]

Katz was "a long-overdue expansion" of Fourth Amendment protection that citizens would need in a technological era.[14] But its "reasonable expectation of privacy" test has exerted little impact on the government's power to conduct technology-aided virtual searches.

The "third party doctrine" of the Fourth Amendment, which is derived from the assumption-of-risk doctrine,[15] facilitates the government's information gathering while end-running the Fourth Amendment's restraints on the government. The third party doctrine posits that "a person does not have a reasonable expectation of privacy in information that he or she has voluntarily exposed or communicated to a third party."[16]

In applying the third party doctrine, the Supreme Court has interpreted the reasonable expectation of privacy so narrowly that it has lost its practical value in privacy jurisprudence. The Court's view of privacy is based on its "secrecy paradigm": privacy exists when information is "completely hidden," and no privacy exists when information is exposed.[17]

This antiquated notion of privacy in the digital era is troubling to some. "Unless we rethink the binary notion of privacy, new technologies will increasingly invade the enclaves of privacy we enjoy in public," the leading privacy law expert Daniel J. Solove argued.[18]

Justice Sonia Sotomayor of the U.S. Supreme Court agrees. In her concurring opinion in *United States v. Jones*,[19] she criticized the third party doctrine for being "ill suited to the digital age, in which people reveal a great deal of information about themselves to third parties in the course of carrying out mundane tasks."[20] Courts should no longer "treat secrecy as a prerequisite for privacy" under the Fourth Amendment, Justice Sotomayor argued.

Given that law tends to react rather than pro-act to new developments in society, it is more challenging now for law to catch up with privacy being reshaped by technology. Some federal statutes are outdated. The Stored Communications Act (SCA), for example, mandates the government to obtain a warrant in searching stored emails and voice messages. But a warrant is not required for access to files stored for six months or more, or stored on a third party's server.

A coalition called "Digital Due Process" is trying to enact the Electronic Communications Privacy Act (ECPA), which includes the SCA, updated and revised. Digital Due Process states: "ECPA has been outpaced. The statute has not undergone a significant revision since it was enacted in 1986—light years ago in Internet time."[21] In September of 2012, an ECPA reform bill was introduced to the U.S. Senate that would require law enforcement officers to obtain a warrant before accessing old private emails and other online communications.

THE FOURTH AMENDMENT IN THE DIGITAL ERA

The U.S. Supreme Court thus far has declined to decide whether text messages should be protected by the Fourth Amendment. In *City of Ontario v. Quon*,[22] the Supreme Court held that Police Sergeant Jeff Quon's text messages that were sent from his police department-issued pager could be searched by the Ontario

(California) Police Department. Even assuming that he had a reasonable expectation of privacy in his text messages, the Court stated that the police department's warrantless review of his pager transcripts was "reasonable."

Unlike the U.S. Supreme Court, a couple of lower federal and state courts have ruled on Fourth Amendment protections of emails and text messages. The Sixth U.S. Circuit Court of Appeals held in *United States v. Warshak*[23] that subscribers enjoy a reasonable expectation of privacy in the email messages they store with an ISP, and thus the government cannot compel the ISP to disclose the subscribers' emails without a judicial warrant based on probable cause.[24]

Warshak started when Steven Warshak challenged the government's warrantless seizure of his emails. Noting that he "plainly manifested" a "subjective expectation" of privacy that his emails would not be disclosed to others, the Sixth Circuit pointed out the "often sensitive and sometimes damning substance" of his emails. "(P)eople seldom unfurl their dirty laundry in plain view," the court said.[25]

On whether society was prepared to recognize as reasonable Warshak's expectation of privacy in the contents of his emails, the Sixth Circuit termed the question one "of grave import and enduring consequence" because email plays a "prominent role" in modern communication.[26]

Regarding the Fourth Amendment protection of emails, the Sixth Circuit took special note of the ISP's role in shielding or releasing the emails. The court emphasized two "bedrock principles":

- First, the very fact that information is being passed through a communications network is a paramount Fourth Amendment consideration.
- Second, the Fourth Amendment must keep pace with the inexorable march of technological progress, or its guarantees will wither and perish.[27]

Likening an email to a postal mail (i.e., snail mail) or a phone call, the Sixth Circuit said an ISP is functionally no different from a post office or a telephone company. The Fourth Amendment prohibits the police from entering a post office to intercept a letter or from using a phone system to secretly record a telephone call without a warrant. Logically, the Fourth Amendment prohibits government agents from compelling an ISP to surrender a subscriber's emails without a warrant.[28] The Sixth Circuit also rejected the third party doctrine, for neither "the mere ability" of an ISP nor its "right" to access its users' emails is sufficient to extinguish or diminish the user's reasonable expectation of privacy.[29]

More recently, the Fourth Amendment protection was at issue when a criminal defendant, Michael Patino, challenged warrantless searches and seizures of his private text messages. Superior Court Judge Judith Savage of Rhode Island ruled in *Rhode Island v. Patino*: "(I)t is objectively reasonable for people to expect the contents of their electronic text messages to remain private, especially vis-à-vis law enforcement."[30]

In applying the two-prong test of *Katz* for privacy (subjective expectation of privacy and objectively reasonable recognition), Judge Savage found that Patino had a "subjective expectation of privacy" in the contents of his text messages because "there is no danger" that no one other than the police saw or overheard his text message.[31] She also determined Patino's expectation of privacy in his text messages to be objectively reasonable. She discounted the "risk" that someone else, not the intended recipient, would receive a text message as "simply too remote to eliminate a person's objectively reasonable belief" that its intended recipient would view the message.[32]

Does the third party doctrine not allow finding an expectation of privacy in the contents of text messages? "If applied absolutely, the third party doctrine would effectively defeat any expectation of privacy in text messages and, potentially, all electronic communications," Judge Savage answered. "This result is untenable, however, in our modern world where electronic communication is omnipresent and a cultural necessity."[33] For text messaging as a mode of communication, the third party doctrine poses a distinct problem because each text message is exposed to several third parties. The "simple technological reality" of text messaging, Judge Savage stated, "should not be allowed to entirely negate an individual's right of privacy."[34]

Judge Savage's opinion was "a bold and interesting one," given that the Fourth Amendment law on text messages is still evolving.[35] It is not likely to be the final word on this unsettled law, for the state of Rhode Island is appealing the court ruling to the Supreme Court of Rhode Island.[36]

PRIVACY IN WORKPLACE EMAILS

Although email is essential as a business tool, it is not necessarily used exclusively for business in the workplace. Employers monitor their employees' email use at work. This is when the employee's right of email privacy collides with the employer's interests in avoiding legal liability, protecting company assets, and preventing loss of productivity.[37]

In assessing an employee's email privacy in the workplace, courts borrow the reasonable expectation of privacy standard from Fourth Amendment law. Meanwhile, an employer's monitoring of an employee's emails can be an invasion of privacy tort in common law. Like in Fourth Amendment law, the reasonableness of a worker's claim for intrusion is composed of subjective and objective elements. As a U.S. bankruptcy court stated in a threshold workplace email case, "one claiming an 'intrusion on seclusion' must show, inter alia, a subjective expectation of privacy and that the expectation is objectively reasonable."[38] And the employee's expectation of privacy hinges on the "operational realities" of the workplace, and thus it "may be reduced by virtue of actual office practices and procedures, or by legitimate regulation."[39]

The U.S. Bankruptcy Court in *Asia Global Crossing Ltd.* laid out a four-part test for an employee's reasonable expectation of privacy relating to email sent and received over a computer owned by a company server. The often-used test states:

- Does the corporation maintain a policy banning personal or other objectionable use?;
- Does the company monitor the use of the employee's computer or e-mail?;
- Do third parties have a right of access to the computer or e-mails?; and
- Did the corporation notify the employee, or was the employee aware, of the use and monitoring policies?[40]

The Supreme Court of New Jersey measured the *Asia Global* factors in *Stengart v. Loving Care Agency*.[41] Marina Stengart sued her ex-employer, Loving Care Agency, after emails with her attorney were discovered by Loving Care in connection with her discrimination lawsuit.[42]

In answering to what extent an employee in Stengart's situation can expect privacy in personal emails, the New Jersey Supreme Court held that "under the circumstances," the employee could "reasonably" expect that the lawyer-client email communications through her personal account would remain private and that sending and receiving emails via a company laptop did not eliminate the lawyer-client privilege that protected them.[43]

Two principal issues for the court were (1) how adequate was Loving Care's electronic communications policy as a notice to its employees? and (2) what were the "public policy concerns" raised by the attorney-client privilege?

Loving Care's policy stipulated its right to review "all matters" on its "media systems and services at any time" and the email messages were considered the company's "business … records."[44] The New Jersey Supreme Court found, however, that the company's policy was unclear on the use of personal password-protected webmail via a company computer. Hence, the court held that employees had no "express notice" that their messages on a personal, web-based email account would be monitored if the account was accessible through company equipment.[45]

Further, Loving Care's policy provided no warning to employees that the contents of their emails were stored on a hard drive and could be forensically recovered by their company. The policy was ambiguous, the New Jersey court continued, since it permitted "occasional personal use" of email via a company computer while declaring that emails were neither private nor personal to individual employees.[46]

The court said "a clear company policy" forbidding personal emails could reduce a possibility that an employee can claim a reasonable expectation of privacy in email messages. Nonetheless, the court was unsure about a company's "zero-tolerance policy," which it found to be impractical in today's mobile workforce.[47]

Meanwhile, the New Jersey Supreme Court ruled that Stengart's email exchanges with her attorney were covered by the attorney-client privilege that promotes the client's "free and full disclosure of information" to the attorney.[48] For her emails to her lawyer related to her working conditions and her contemplated lawsuit against Loving Care.

The California Court of Appeal ruled in January 2011 that the attorney-client privilege did not protect the emails that were sent via a company email account. In *Holmes v. Petrovich Development Co.*,[49] Gina Holmes, a clerk, sued a small-business employer over, among other things, sexual harassment, retaliation, and violation of the right of privacy. Her former employer introduced her email messages with her attorney to challenge the veracity of her claims. On appeal, Holmes argued that the trial court erred in allowing the emails into the case.[50]

In upholding the trial court's ruling against Holmes, the California appellate court stated that her emailing through her company computer was not private. The court reasoned that her email exchanges were "akin to consulting her attorney in one of (her employer's) conference rooms, in a loud voice, with the door open, yet unreasonably expecting that her conversation that her employer overheard would be 'privileged.'"[51]

In 2011, a U.S. district court focused on the email policy of a money market fund when the Securities and Exchange Commission (SEC) sought access to the emails of the fund's vice chairman. *In re Reserve Fund Securities and Derivative Litigation*[52] stemmed from the SEC request for the emails that Bruce Bent II, the vice chairman of the Reserve Management Co., sent to his wife through his company computer from his company email account. Bent claimed that the emails were protected by his "marital privilege."[53]

Judge Paul G. Gardephe rejected Bent's claim. He first noted the explicit company policy of limiting email use to official business and directing employees to delete personal emails from inboxes "on a regular basis."[54] He then mentioned the email policy that the employees' emails were automatically saved "regardless of content" and subject to "regulatory agencies or the courts."[55] In addition, the company reserved the right to access the employees' emails "for a legitimate business reason."[56]

In ruling against Bent, Judge Gardephe applied the balancing test of Asia Global for email privacy. Because Bent had sent his emails to his wife through his company email system, Judge Gardephe held, they were not sent "in confidence" and thus the "marital communications privilege" could not protect them.[57]

A Massachusetts trial court in 2011 ruled against a firefighter in Falmouth in an email privacy lawsuit. In *Falmouth Fire Fighters' Union Local 1497 v. Town of Falmouth*,[58] the president of the Fire Fighters' Union claimed that the

Massachusetts Privacy Act had been violated when the town of Falmouth accessed and disseminated his emails in connection with a police investigation and lawsuit.[59]

Judge Christopher J. Muse of the Massachusetts Superior Court differed with the fire fighters' union president, Russell Ferreira, that he had an expectation of privacy in his emails. Judge Muse said Ferreira "voluntarily" sent his emails over the town's email system although he had no assurance that his emails would be "private and confidential."[60] Judge Muse added that once "highly personal and intimate" emails were sent to a second person, "any reasonable expectation of privacy" was lost.[61]

EMAIL AND FREEDOM OF INFORMATION

Access to government records is not a constitutional right in the United States. Rather, it is a statutory right. The federal Freedom of Information Act (FOIA) was passed by Congress in 1966 to allow individuals access to government agency records for public inspection. Disclosure, not secrecy, was the primary objective of the FOIA although it provided for nine exemptions.

The FOIA was amended in 1996 to facilitate access to electronic records. The Electronic FOIA required government agencies to index their records online and to provide records in electronic form to those who want that format if the records are available electronically. In order to address the growing post-9/11 secrecy issues, Congress in 2007 enacted the Openness Promotes Effectiveness in our National (OPEN) Government Act. The OPEN Government Act established the Office of Government Information Services (OGIS) and a FOIA ombudsman. The law also made the FOIA deadlines more consequential; agencies that fail to respond to FOIA requests within the deadline cannot charge search fees.

Every state has its own FOI law. Some states such as Florida protect access to government information as a constitutional right. A number of state open records laws have been patterned after the federal FOIA. So, the state FOI statutes are similar to the federal FOIA.

Email in FOI Statutes: Federal and State

The federal FOIA applies to "any information that would be an agency record" subject to the FOIA requirements maintained by an agency "in any format, including an electronic format."[62] Few FOIA cases have arisen from the format of the requested government information as a central issue. Regardless, one FOI authority, emphasizing the "new privacy concerns" over public officials' increasing use of emails, commented:

> While the vast majority of jurisdictions that have addressed the issue have classified e-mails created by or stored on public computers as public records, some exclude personal e-mails

from the scope of public records laws. Other jurisdictions distinguish between the content of e-mails and the e-mail addresses of those sending and receiving the messages. Some jurisdictions have sought to balance the public interest in access to public employees' e-mail messages with the privacy interests implicated by disclosure.[63]

Among the few states where public records laws include emails "explicitly" in defining a record are California and Colorado.[64] A majority of states consider their FOI laws reaching emails. The Arkansas open records law defines a public record as encompassing "electronic or computer-based information." Likewise, the District of Columbia law states that "public record" covers "information stored in an electronic format."[65]

Is email a public record under FOI laws? Many states treat email as a public record. But at least six jurisdictions have not yet addressed or have been unclear on the issue: Hawaii, Michigan, Nevada, Rhode Island, Vermont, and Wyoming.[66]

Are public matters on government email or government hardware subject to disclosure? A number of states are similar to North Carolina on this FOI issue: "Given the definition of public records as those records related to the transaction of public business, all materials on a government-issued computer or email address should be public if they relate to government business."[67] Neither statutory nor case law in 10 jurisdictions has addressed this question specifically.[68]

Do FOI statutes cover a private matter on government email or government hardware? More often than not, if an email has little to do with public business, it is considered not to be a public record. New Hampshire, New Mexico, North Carolina, and Oregon are illustrative.[69] But other states such as Georgia are more access-friendly unless the emails are not specifically exempted.[70] More than a dozen states and the District of Columbia do not have statutory or case law on this issue.[71]

What is the FOI status on the public matter on private email? While 26 states consider private emails, when used for government business, as public records, other states have "no clear rules or prevailing case law."[72] The Alaska Supreme Court held in 2012 that it is not "a per se violation" of the state's Public Records Act for public employees to use private email accounts in conducting state business. The court also stated that private emails relating to state business are not distinguished from other "public records" that are preserved under the law.[73]

Federal rules require government agencies to address the use of private email systems not controlled by the agency (such as commercial email systems like Gmail and Hotmail). If agency staff use external private email systems, "agencies must ensure that federal records sent or received on such systems are preserved in the appropriate recordkeeping system and that reasonable steps are taken to capture available transmission and receipt data needed by the agency for recordkeeping purposes."[74] There is little FOIA case law on whether the use of private email accounts to discuss official business can trigger the FOIA application to the emails.[75]

Finally, do private matters contained in private emails fall within the scope of FOI law? With few exceptions, emails of purely private or personal nature are not treated as public records. Oregon is typical in not applying FOI statutes to private emails on private computers: "If the email does not relate to the public's business and it is contained on a privately owned computer, then it is by statutory definition not a public record."[76]

Email in FOI Statutes: Judicial Interpretations

Media lawyer Peter S. Kozinets,[77] commenting on access to public officials' emails as a public's right to know, wrote: "(R)ecent court decisions involving access to electronic mail messages on government computer systems have created a distinction between public and personal emails that threatens to erode the strong historic presumption of public access to governmental records."[78] While detecting a judicial trend to separate public from personal emails of government officials, Kozinets concluded that the major state court cases he analyzed "likely reflect an early and unfinished response" to the question of how to balance the public's right to information with the public employees' privacy in personal emails.[79]

For example, in *Easton Area School District v. Baxter*,[80] one of the most recent FOI cases on emails, the Pennsylvania Commonwealth Court held:

> (E)mails should not be considered "records" just because they are sent or received using an agency email address or by virtue of their location on an agency-owned computer, even where (…) the agency has a policy limiting use of computers to official business and stating that users have no expectation of privacy. That is so because a record is "information … that documents a transaction or activity of an agency," and personal emails that do not do so are simply not records.[81]

The Pennsylvania court found that the emails were a record under the state's public records law as "information" that documents "a transaction or activity of an agency."[82]

The *Baxter* court's emphasis on the nature and purpose of the government official's emails, not their location, is hardly surprising. Few other state courts have accepted the contention that "all emails on a government computer are automatically public records."[83]

One of the earliest state supreme court cases on emails was *Ohio ex rel. Wilson-Simmons v. Lake County Sheriff's Department*.[84] The Ohio Supreme Court refused to hold that allegedly racist emails between corrections officers were public records. They were not the kind of documents, under Ohio's Public Records Act, that were created or received by a public office, which "serves to document the organization, functions, policies, decisions, procedures, operations, or other activities of the office," the court said.[85]

The Florida Supreme Court in *Florida v. City of Clearwater*[86] followed suit in 2003: "(P)rivate documents cannot be deemed public records solely by virtue of their placement on an agency-owned computer. The determining factor is the nature of the record, not its physical location."[87]

The Colorado Supreme Court in *Denver Publishing Co. v. Board of County Commissioners of County of Arapahoe*[88] was similar to the Florida Supreme Court. In 2005, the Colorado court held that a public official's mere act of possessing, creating, or receiving an email record does not automatically make the email a public record, because the email must be demonstrably connected to the performance of public functions or to the receiving or spending of public funds.[89] The emails requested by a media company under the Colorado Open Records Act were "sexually-explicit" exchanges between two government officials, and they were for their "personal relationship" and not in the performance of their official duties.[90]

In finding an email on a public computer system to be a public record for disclosure, the Supreme Courts in Arizona, Arkansas, and Idaho have agreed on the need for a nexus between the email and the public's business. In *Griffis v. Pinal County*,[91] the Arizona Supreme Court said the "public records" definition did not extend to "purely private or personal" documents. Rather, it applied to "only those documents having a 'substantial nexus' with a government agency's activities."[92] The "nature and purpose" of the records determined their public status, which required "a content-driven inquiry."[93]

The Idaho Supreme Court in *Cowles Publishing Co. v. Kootenai County Board of County Commissioners*[94] ruled that the emails between a manager of the juvenile education and training court (JETC) and her supervisor, the county prosecutor, were public records under Idaho's Public Records Act. The Idaho Supreme Court held: "It is not simply the fact that the emails were sent and received while the employees were at work ... that makes them a public record. Rather, it is their relation to legitimate public interest that makes them a public record."[95] The court focused more on the context of the emails than on their contents.

The Arkansas Supreme Court in *Pulaski County v. Arkansas Democrat-Gazette*[96] required "a substantial nexus" between emails and a government agency's activities. "Comparing the nature and purpose of a document with an official's or agency's activities to determine whether the required nexus exists necessarily requires a fact-specific inquiry," the *Arkansas* court held.[97] The Arkansas Supreme Court's approach was more content-oriented than context-based.

The West Virginia Supreme Court of Appeals in *Associated Press v. Canterbury*[98] has adopted the "majority position" on whether personal emails were subject to disclosure under FOI laws. The court held that the emails between a judge and the CEO of an energy company were not a "public record" under the Freedom of Information Act of West Virginia. Noting the decisions from six other states for support, the highest court of West Virginia stated: "(T)he majority of courts have held that personal email communication by a public official or employee does not constitute a public record for purposes of disclosure under FOIA or its equivalent."⁹⁹

The *West Virginia* court said its holding was "consistent" with federal courts' interpretations of the federal FOIA.¹⁰⁰ The *West Virginia* court viewed the content analysis as a "better approach" in determining whether an email is a public record under its state law.¹⁰¹

In 2010, the Wisconsin Supreme Court rejected an FOI request for all emails that public school teachers sent and received via their school district email accounts on school district-owned computers. In *Schill v. Wisconsin Rapids School District*,¹⁰² the court ruled: "While government business is to be kept open, the contents of employees' personal emails are not a part of government business ... simply because they are sent and received on government email and computer systems."¹⁰³

State FOI case law indicates that the balancing between disclosure of emails and privacy has often been struck in favor of non-disclosure to protect privacy. One of the rare exceptions to the non-disclosure approach in FOI law was a California appellate court case of 2003. Although it was "unpublished,"¹⁰⁴ the California Court of Appeals' opinion in *Holman v. Superior Court of San Diego County*¹⁰⁵ is noteworthy for it is detailed and comprehensive in addressing the privacy issues enveloping emails to and from government officials.

Holman arose from the city of San Diego's rejection of the *San Diego Reader*'s request under the California Public Records Act (CPRA) for "copies of all emails" to a former press secretary for the mayor of San Diego.¹⁰⁶ The trial court agreed with the city in denying the media's FOI request.

The California Court of Appeal disagreed with the trial court on the privacy in the email exchanges at issue. Opting for a balancing test under California's public records law, the court declared: "(N)o absolute privilege" denies access to private identifying information unless it is protected by the constitutional right of privacy in California and thus exempted from disclosure.¹⁰⁷

The California court examined the balancing of disclosure vs. privacy under the California public records law. The court first focused on the extent to which the disclosure of the emails would violate the asserted privacy concerns. To answer this question, judges should examine whether there is a "legally protected" privacy interest and whether there is a "reasonable expectation of privacy" under the circumstances.¹⁰⁸

Noting numerous courts that have dealt with the email privacy issues, the *Holman* court stated that those using email for communication have no reasonable expectation of privacy in their emails and that the revised Public Records Act left no doubt that the emails retained by local government agencies become government records.¹⁰⁹ The court added that releasing the content of the emails without revealing the correspondent's email addresses would violate no protected privacy interest of the parties involved.¹¹⁰

On the "chilling effect" of disclosure of the email addresses on the public and the media, the California Courts of Appeal diametrically disagreed with the trial court. When compared with the "disclosure interest" of the scale, i.e., accessing vital information about the city government's conduct of its business, the California appellate court determined the "non-disclosure interest," i.e., privacy protection, to be "minimal."[111]

Six years later, access to email addresses in a government agency record was in dispute in an FOI case in the state of Washington. The email addresses concerned elected officials, not those who emailed the government officials. Meredith Mechling requested email messages of the city council members who discussed city business in Monroe. The trial court decided that the council members' personal email addresses were exempt from disclosure under Washington's Public Disclosure Act. The Washington Court of Appeals in *Mechling v. City of Monroe*[112] reversed the trial court. In interpreting the personal information exemption under the public records law, the Washington appellate court held: "(T)he exemption only applies to the personal email addresses contained in personnel or employment related records held by the public agency and does not exempt from disclosure personal email addresses contained in the email messages of public officials discussing City business."[113]

Regarding the attorney-client privilege to the city council members' email messages, the Washington Court of Appeals applied the privilege narrowly. Because it is confined to information relating to attorney advice, the court held, the privilege does not extend to documents prepared for unprivileged communication with an attorney.[114]

Are emails contained on a government official's personal computer public records? That was the question before the Pennsylvania Commonwealth Court in *In re Silberstein*,[115] a case of first impression under Pennsylvania's new Right-to-Know Law. The court decided that the emails on York Township commissioner Kenneth M. Silberstein were not public records as defined by the law. The court distinguished "transactions or activities of an agency" that might be a public record from the "emails or documents of an individual public office holder" that might not be a public record.[116] Since Silberstein, an individual public official, had no authority to act alone on behalf of the Township, the emails on his personal computer would not be public records, insofar as they were personally and individually created by him and did not document a transaction or activity of York Township as the local agency.[117]

SUMMARY AND CONCLUSIONS

New technology, especially when it develops breathtakingly fast, will challenge and change people's conception of privacy and their "reasonable expectation" of privacy. Social understanding and technology are forcing an unending societal rebalancing of privacy in Internet communication with governmental and individual interests.

The legislative and judicial approach to the Fourth Amendment and common law on privacy illustrates how the law on the books should be retooled to accommodate the law in action. The shift from place to people as the central focus of privacy under the Fourth Amendment in the mid-1960s was literally seismic although it came rather belatedly. On the other hand, the practical impact of the reasonable expectation of privacy on Americans for more protection was less than expected. The third party doctrine has been resorted to by the government to more or less neutralize the subjective and objective reasonableness of privacy. It is singularly important that some judges, including Justice Sotomayor of the U.S. Supreme Court, are now questioning the third party doctrine.

Although the U.S. Supreme Court hesitates to address privacy in the digital century, lower federal and state courts are noticeably enterprising in reinterpreting the Fourth Amendment. In a way, they are willing to "think and live outside the box" in making privacy a living concept. Their broad applications of the Fourth Amendment's restraints in protection of privacy in emails and text messages are in no small measure a subtle but unmistakable repudiation of the U.S. Supreme Court's largely pre-Internet interpretations of privacy that are anchored to the assumption-of-risk argument.

Email privacy in the workplace reflects the Fourth Amendment's benchmark (i.e., "reasonable expectation of privacy"), and its protection or non-protection has more to do with the workplace's "operational realities" than anything else. The 4-prong test of *Asia Global* epitomizes the fact-specific balancing approach of American courts as a whole, when employees take few proactive actions on their email communications such as using personal, password-protected web emails.

Government employees' privacy in emails under federal and state FOI laws depends on how public records are statutorily defined. Like in the workplace email cases, the reasonable expectation of privacy informs courts in deciding whether the "personal" emails of government officials should be personal and thus not be disclosed to the public. The majority of court cases have examined what the requested emails concerned, not where they were created and stored. That is, the content of the emails, within the context of their close nexus to government business, not their location, matters. Disclosure of personal emails due to their location in government offices is the exception, not the rule.

Citizens' expectation of privacy in their emails to government agencies and government officials has not been often adjudicated. It should not be a major issue. Emailing tends to be intrinsically not as private as it is assumed, and the inhibiting impact of email disclosure on potential individual communicators in relation to government business ought not to be overestimated.

NOTES

1. Matthew Collins, *The Law of Defamation and the Internet* (3rd ed.). (Oxford: Oxford University Press, 2010): 23.
2. Radicati Group, "Email Statistics Report, 2010–2014," http://www.radicati.com
3. *Garrity v. John Hancock Mutual Life Insurance Co.*, 2002 WL 974676 (D. Mass. 2002).
4. Marc A. Sherman, "Webmail at Work: The Case for Protection Against Employer Monitoring," *Touro Law Review* 23 (2007): 647–684.
5. Claire Cain Miller, "Should E-mail and Letters Have Equal Legal Protection?" *New York Times*, January 13, 2011, http://bits.blogs.nytimes.com/2011/01/13/should-e-mail-and-letters-have-equal-protection/
6. Richard J. Peltz-Steele, *The Law of Access to Government* (Durham, NC: Carolina Academic Press, 2012): 400.
7. Stephen Braun, "Mitt Romney Used Private Email Accounts to Conduct State Business While Massachusetts Governor," *Huffington Post*, March 9, 2012, http://www.huffingtonpost.com/2012/03/09/mitt-romney-emails_n_1335712.html
8. *Competitive Enterprise Institute v. United States Environmental Protection Agency*, No. 1:12-cv-01617 (D.D.C. 2012), http://www.scribd.com/doc/108554821/CEI-v-EPA-Complaint-Re-Secret-Accounts
9. Kyu Ho Youm, "International and Foreign Law," In *Media Law and Ethics* (4th ed.), Roy L. Moore and Michael D. Murray, (Eds.) (New York, NY: Routledge, 2012): 686.
10. Akhil R. Amar, *America's Unwritten Constitution: The Precedents and Principles We Live By* (New York, NY: Basic Books, 2012): 126.
11. *Katz v. United States*, 389 U.S. 347 (1967).
12. *Katz v. United States*, 351.
13. *Katz v. United States*, 361.
14. Christopher Slobogin, "Is the Fourth Amendment Relevant in a Technological Age?" In Jeffrey Rosen and Benjamin Wittes (eds.), *Constitution 3.0: Freedom and Technological Change* (Washington, DC: Brookings Institution Press, 2011): 13.
15. The assumption-of-risk doctrine allows little reasonable expectation of privacy for people who disclose information to third parties and thus assume the risk that the information will be made available to the government. See Slobogin, "Is the Fourth Amendment Relevant," 17.
16. *Rhode Island v. Patino*, 2012 R.I. Super. LEXIS 139 (Sup. Ct. R.I. 2012), 72.
17. Daniel J. Solove, 2012. "*United States v. Jones* and the Future of Privacy Law: The Potential Far-Reaching Implications of the GPS Surveillance Case," *Privacy and Security Report*, January 30, 2012, 3.
18. Daniel J. Solove, *The Future of Reputation: Gossip, Rumor, and Privacy on the Internet* (New Haven, CT: Yale University Press, 2007).
19. *United States v. Jones*, 132 S. Ct. 945 (2012).
20. *United States v. Jones*, 957.
21. Digital Due Process, "About the Issue," http://digitaldueprocess.org/index.cfm?objectid=37940370-2551-11DF-8E02000C296BA163
22. *City of Ontario v. Quon*, 130 S. Ct. 2619 (2010).
23. *United States v. Warshak*, 631 F.3d 266 (6th Cir. 2010), rehearing and rehearing en banc denied, March 7, 2011.
24. *United States v. Warshak*, 288.

25. *United States v. Warshak.*
26. *United States v. Warshak.*
27. *United States v. Warshak,* 285.
28. *United States v. Warshak.*
29. *United States v. Warshak,* 286–87.
30. *Rhode Island v. Patino,* 116.
31. *Rhode Island v. Patino,* 113.
32. *Rhode Island v. Patino,* 117.
33. *Rhode Island v. Patino,* 118–19.
34. *Rhode Island v. Patino.*
35. Jon Brodkin, "Police seizure of text messages violated 4th Amendment, judge rules." *Ars Technica,* http://arstechnica.com/tech-policy/2012/09/police-seizure-of-text-messages-violated-4th-amendment-judge-rules/
36. Courtney Caligiuri, "Hearing for Man Accused of Killing Boy," *WPRI.com Eyewitness News,* September 10, 2012.
37. Sherman, "Webmail at Work," 657–60.
38. *In re Asia Global Crossing, Ltd.,* 322 B.R. 247 (Bankr. S.D.N.Y. 2005), 257.
39. *O'Connor v. Ortega,* 480 U.S. 709 (1987), 717.
40. *In re Asia Global Crossing,* 257.
41. *Stengart v. Loving Care Agency,* 990 A.2d 650 (N.J. Sup. Ct. 2010).
42. *Stengart v. Loving Care Agency,* 655.
43. *Stengart v. Loving Care Agency.*
44. *Stengart v. Loving Care Agency,* 659.
45. *Stengart v. Loving Care Agency.*
46. *Stengart v. Loving Care Agency.*
47. *Stengart v. Loving Care Agency,* 663.
48. *Stengart v. Loving Care Agency.*
49. *Holmes v. Petrovich Development Co.,* 191 Cal. App. 4th 1047 (Cal. App. Ct. 2011).
50. *Holmes v. Petrovich Development Co.,* 1068.
51. *Holmes v. Petrovich Development Co.*
52. *In re the Reserve Fund Securities and Derivative Litigation,* 275 F.R.D. 154 (S.D.N.Y. 2011).
53. *In re the Reserve Fund Securities,* 156.
54. *In re the Reserve Fund Securities,* 158.
55. *In re the Reserve Fund Securities,* 159.
56. *In re the Reserve Fund Securities.*
57. *In re the Reserve Fund Securities,* 164.
58. *Falmouth Fire Fighters' Union Local 1497 v. Town of Falmouth,* 2011 WL 7788014 (Mass. Super. Ct. 2011).
59. *Falmouth Fire Fighters' Union Local 1497 v. Town of Falmouth.*
60. *Falmouth Fire Fighters' Union Local 1497 v. Town of Falmouth.*
61. *Falmouth Fire Fighters' Union Local 1497 v. Town of Falmouth.*
62. *Freedom of Information Act,* 5 U.S.C. §§552(b)(1)–(b)(9), 1966.
63. Lee Levine, Robert C. Lind, Seth Berlin, and C. Thomas Dienes, *Newsgathering and the Law* (4th ed.) (Charlottesville, VA: Matthew Bender & Co., 2011).

64. The Reporters Committee for Freedom of the Press website, "Open Government Guide," is the most comprehensive resource on state FOI laws: http://www.rcfp.org/open-government-guide. The author's discussion of state FOI is substantially based on this RCFP data.
65. Reporters Committee, "Open Government Guide."
66. Reporters Committee, "Open Government Guide."
67. Reporters Committee, "Open Government Guide."
68. Reporters Committee, "Open Government Guide."
69. Reporters Committee, "Open Government Guide."
70. Reporters Committee, "Open Government Guide."
71. Reporters Committee, "Open Government Guide."
72. Braun, "Mitt Romney Used Private Email Accounts."
73. *McLeod v. Parnell*, 2012 WL 4840769 (Alaska Sup. Ct. 2012).
74. Government Accountability Office, "Federal Records: National Archives and Selected Agencies Need to Strengthen E-mail Management," GAO-08-742, 2008, http://www.gao.gov/assets/280/276561.pdf
75. Michael D. Pepson and Daniel Z. Epstein, "Gmail.gov: When Politics Gets Personal, Does the Public Have a Right to Know?" *Engage*, 13 (July 2012): 4–16.
76. Pepson and Epstein, "Gmail.gov."
77. Peter S. Kozinets, "Access to the E-mail Records of Public Officials: Safeguarding the Public's Right to Know," *Communications Lawyer*, 15 (summer 2007): 17–26.
78. Kozinets, "Access to E-mail Records of Public Officials," 17.
79. Kozinets, "Access to E-mail Records of Public Officials," 24.
80. *Easton Area School District v. Baxter*, 35 A.3d 1259 (Pa. Commw. Ct. 2012).
81. *Easton Area School District v. Baxter*, 1264.
82. *Easton Area School District v. Baxter*.
83. *Easton Area School District v. Baxter*, 1263.
84. *Ohio ex rel. Wilson-Simmons v. Lake County Sheriff's Department*, 693 N.E.2d 789 (Ohio 1998), reconsideration denied, 696 N.E.2d 226 (Ohio 1998).
85. *Ohio ex rel. Wilson-Simmons v. Lake County Sheriff's Department*, 793.
86. *Florida v. City of Clearwater*, 863 So. 2d 149 (Fla. 2003).
87. *Florida v. City of Clearwater*, 154.
88. *Denver Publishing Co. v. Board of Colorado Commissioners of Arapahoe*, 121 P.3d 190 (Colo. 2005).
89. *Denver Publishing Co. v. Board of Colorado Commissioners of Arapahoe*, 203.
90. *Denver Publishing Co. v. Board of Colorado Commissioners of Arapahoe*.
91. *Griffis v. Pinal County*, 156 P.3d 418 (Ariz. 2007).
92. *Griffis v. Pinal County*, 421.
93. *Griffis v. Pinal County*.
94. *Cowles Publishing Co. v. Kootenai County Board of County Commissioners*, 159 P.3d 896 (Idaho 2007).
95. *Cowles Publishing Co. v. Kootenai County Board of County Commissioners*, 901.
96. *Pulaski County v. Arkansas Democrat-Gazette, Inc.*, 260 S.W.3d 718 (Ark. 2007).
97. *Pulaski County v. Arkansas Democrat-Gazette, Inc.*, 724.
98. *Associated Press v. Canterbury*, 688 S.E.2d 317 (W. Va. 2009).
99. *Associated Press v. Canterbury*, 325.

100. *Associated Press v. Canterbury*, 331.
101. *Associated Press v. Canterbury*, 335.
102. *Schill v. Wisconsin Rapids School District*, 786 N.W.2d 177 (Wisc. 2010).
103. *Schill v. Wisconsin Rapids School District*, 183.
104. An "unpublished" opinion refers to "(a)n opinion that the court has specifically designated as not for publication." See Bryan A. Garner (ed.), *Black's Law Dictionary* (St. Paul, MN: Thomson Reuters, 2009): 1202. *Black's Law Dictionary* notes: "Court rules usually prohibit citing an unpublished opinion as authority. Such an opinion is considered binding only on the parties to the particular case in which it is issued."
105. *Holman v. Superior Court of San Diego County*, 31 Media L. Rep. (BNA) 1993 (Cal. Ct. App. 2003).
106. *Holman v. Superior Court of San Diego County*, 1994–95.
107. *Holman v. Superior Court of San Diego County*, 1999.
108. *Holman v. Superior Court of San Diego County*, 1999–2000.
109. *Holman v. Superior Court of San Diego County*, 2000.
110. *Holman v. Superior Court of San Diego County*.
111. *Holman v. Superior Court of San Diego County*.
112. *Mechling v. City of Monroe*, 222 P.3d 808 (Wash. Ct. App. 2009), review denied, 236 P.3d 206 (Wash. 2010).
113. *Mechling v. City of Monroe*, 846.
114. *Mechling v. City of Monroe*, 853.
115. *In re Silberstein*, 11 A.3d 629 (Pa. Commw. Ct. 2011).
116. *In re Silberstein*, 633.
117. *In re Silberstein*.

PART THREE

Looking Ahead: Leaker Chaos, Resolution, and the Millennial Shift

CHAPTER EIGHT

All the News That's Fit to Leak

JONATHAN PETERS

Alan Rusbridger, editor of *The Guardian*, told members of Parliament in December 2013 that after his newspaper obtained documents on government surveillance from Edward Snowden, a former National Security Agency (NSA) contractor, Rusbridger and his top editors met with government officials in the United States and Britain more than 100 times and were subjected to measures "designed to intimidate."[1] The measures included prior restraints on publication and visits by government officials to Rusbridger's office, as well as "the enforced destruction of *Guardian* computer disks with power tools."[2]

Rusbridger's testimony, before a Parliamentary committee on national security, addressed the delicate balance of free expression and government secrecy, an evergreen issue that became more acute when *The Guardian* began publishing, in June 2013, material leaked by Snowden.[3] The British and American governments have said Snowden's disclosures and the newspaper's reporting, principally detailing NSA efforts to conduct clandestine mass electronic surveillance and data mining, have damaged national security and helped hostile governments.[4] Meanwhile, journalists and transparency advocates have said the disclosures and reporting "spurred a vital debate on privacy and the role of spy agencies in the Internet age."[5]

The ease of anyone—a government employee, foreign national, citizen, or journalist—uploading government data quickly to the Internet for the world to see has raised new issues that are embroiling the world of government access and individual privacy. Revelations of the U.S. government spying on other nations, its own citizens, and journalists—and gathering massive amounts of data about individuals—has further complicated the terrain.

News organizations, especially those that have reported on material leaked by Snowden, have had "to adjust to a harsh new reporting environment," as governments and others have expressed intense interest in the classified material held by the news organizations and their reporters.[6] For them, the risk is twofold: (1) a technological risk that some party might access their data without their knowledge, and (2) a legal risk that some party might use a subpoena or other legal writ to access their data with or without their knowledge. With that in mind, editors have enhanced their internal security arrangements. Marc Frons, chief information officer of the *New York Times*, said, "The old model was kind of like your house. You locked your front door and windows but not your desk drawer, even if it had your passport inside. In the new model, you have locks on everything."[7]

The Guardian went to great lengths to protect the material leaked by Snowden, taking more security precautions and using more countermeasures than ever before for any story.[8] Senior editors turned in their cellphones before discussing the material. They met in rooms without windows and unplugged any nearby electronic devices. Computers storing the material were disconnected from the Internet. Reporters who needed to consult with colleagues in other countries actually flew to those countries, sparing no expense, to meet in person to discuss the material.[9]

This may not be the new normal, because Snowden-scale leaks are rare, but in general the social power of technology is reshaping the gathering, production, and distribution of news and information. Paradoxically, whistleblowers and journalists committed to divulging secrets increasingly depend on technologies and practices designed to keep secrets.[10] Of course, the drive to expose—to blow the whistle or to air a superior's dirty laundry—always has existed.[11] But the technology enabling such exposure, since the dawn of the Internet, has "entered a Cambrian explosion, replicating its effective features, excising its failed components, and honing its methods faster than ever before."[12]

As tech reporter Andy Greenberg concluded in his book *This Machine Kills Secrets*, today the state of the world's information favors the leaker more than ever.[13] Consider the stock worldwide of digitally recorded data. In 2002, the amount matched that of analog-recorded data.[14] Jump ahead only five years, to 2007, and the amount of digitally recorded data made up 94 percent of the world's recorded information.[15] Importantly, all of those digitally recorded data are liquid—readily reproducible, highly mobile, and easily leakable.[16] Many barriers to leaking, in particular megaleaking, have crumbled.[17] It's no longer necessary to gather tens of thousands of paper documents and spend a year photocopying.[18] Nor is it necessary to be a war hero who reaches the government's highest levels only to go rogue.[19] Rather, a leaker could be one of the millions of Americans who has access to secret government data or one of the countless millions of Americans who has access to secret corporate data.[20]

Indeed, roughly 4 million Americans have some form of clearance to access classified information, and of them roughly 1.2 million have "top secret" clearance.[21] Peter Galison, the Pellegrino University Professor in History of Science and Physics at Harvard University recently estimated that "there are five times as many pages being added to the world's classified libraries as to its unclassified ones."[22] This is remarkable in part because of President Barack Obama's memorandum in 2009 stating his commitment to create "an unprecedented level of openness in Government"[23] and his comment in 2013 that his has been "the most transparent administration in history."[24] Obama also said in 2009 that whistleblowing is an act "of courage and patriotism, which can sometimes save lives and often taxpayer dollars" and "should be encouraged rather than stifled."[25] The numbers do not support those claims.

'WORSE THAN NIXON'

In 2010, 76.7 million documents were classified, compared with 8.6 million in 2001 and 23.4 million in 2008, the first and last years of President George W. Bush's administration.[26] Further, Obama has used espionage charges to prosecute more people as whistleblowers than all former presidents combined.[27] They include Jeffrey Sterling, an ex-CIA analyst who gave classified information to Pulitzer Prize-winning reporter James Risen about the agency's botched attempt to sabotage Iran's nuclear-development plans.[28] FBI translator Shamai Leibowitz pleaded guilty to the unlawful disclosure of classified transcripts of bugged conversations in the Israeli embassy; he disclosed them to a blog with the intent to stem Israeli aggression toward Iran.[29] Stephen J. Kim, a senior analyst and arms expert for the State Department and the Lawrence Livermore National Laboratory, was prosecuted for leaking a classified report to Fox News about North Korea's plans to develop nuclear weapons.[30] Ex-CIA officer John Kiriakou pleaded guilty to the unlawful disclosure of classified information about a covert CIA operative; he leaked it to various news outlets, including the *New York Times*.[31] Chelsea Manning, a former Army intelligence analyst, was convicted of espionage after she was identified as the source for WikiLeaks' biggest document dumps.[32] And, finally, Snowden himself faces espionage charges.[33]

In several of those cases, "government officials seized journalists' phone and email records to use in their investigations."[34] For example, in May 2013, the Department of Justice (DOJ) confirmed that it had obtained a "portfolio of information" on *Fox News* correspondent James Rosen in connection with the government's investigation and prosecution of Stephen J. Kim.[35] The DOJ characterized

Rosen as a criminal co-conspirator in order to obtain the "portfolio," which included personal emails and records of Rosen's visits to the State Department.[36] Michael Clemente, *Fox News* executive vice president, said he was "outraged" and that *Fox* would "defend [Rosen's] right to operate as a member of what up until now has always been a free press."[37]

Criticism of the DOJ was swift and harsh, and it came from all directions. Lucy Dalglish, dean of the University of Maryland's Philip Merrill College of Journalism, commented on the chilling effect created by the government's aggressive tactics: "The message is loud and clear that if you work for the federal government and talk to a reporter that we will find you."[38] More broadly, James Goodale, who was general counsel of the *New York Times* during the Pentagon Papers coverage and case, said Obama's approach to classified information and press freedom has been "antediluvian, conservative, backwards" and "worse than Nixon."[39] Echoing those sentiments, Leonard Downie, Jr., former editor of the *Washington Post*, produced in October 2013 a report for the Committee to Protect Journalists that concluded: "The administration's war on leaks and other efforts to control information are the most aggressive I've seen since the Nixon administration. …The 30 experienced Washington journalists at a variety of news publications whom I interviewed for this report could not remember any precedent."[40]

Against that backdrop, the general problem is that the White House "curbs routine disclosure of information" and uses "aggressive prosecution of leakers of classified information and broad electronic surveillance programs [to] deter government sources from speaking to journalists."[41] With regard to the electronic surveillance programs, news stories based on the Snowden material, which revealed extensive surveillance of Americans' telephone and e-mail traffic, have complicated the newsgathering process. Government officials are reluctant to discuss even unclassified information with journalists because the officials fear that leak investigations and surveillance have made it more difficult for journalists to protect them as sources.[42]

The Snowden material has revealed "details of secret NSA operations that acquire, store, and search huge amounts of telephone call, text, and e-mail data from American telephone and Internet companies," all "to find and track communications that might be tied to terrorist activity."[43] It is worth mentioning, too, that *Wired* reported in May 2012 that the NSA has been building the country's biggest spy center to store and process all manner of data.[44] The physical complex, located in Bluffdale, Utah, is so large that it necessitated expanding the town's boundaries; it is more than five times the size of the U.S. Capitol.[45] According to the *Wired* story:

> Flowing through its servers and routers and stored in near-bottomless databases will be all forms of communication, including the complete contents of private emails, cell phone calls, and Google searches, as well as all sorts of personal data trails—parking receipts, travel itineraries, bookstore purchases, and other digital "pocket litter."[46]

The data stored at the Utah complex will go far beyond the world's billions of public web pages.[47] The NSA is more interested in what is called the "invisible web," also known as the "deep web" or "deepnet"—data beyond the public's reach.[48] This includes password-protected data, U.S. and foreign government communications, and noncommercial file-sharing between trusted peers.[49] As of this writing, the complex had not opened because of chronic electrical surges that destroyed hundreds of thousands of dollars worth of computer equipment.

INDIVIDUAL FREEDOMS AND THE SPIRIT OF DEMOCRACY

So far no direct connection has been made between the electronic surveillance programs and the leak investigations, but they are related insofar as the surveillance has added to the "fearful atmosphere" that is enveloping "contacts between American journalists and government sources."[50] The surveillance is significant, too, because many journalists and news outlets have been early adopters of cloud technology.[51] Consider these examples, two involving Amazon's Elastic Compute Cloud (EC2) service, which makes web-scale computing easier for developers, and a third involving Google's corporate email service.

First, in 2008, during the Democratic presidential primary, an engineer at the *Washington Post* used the EC2 to process Hillary Clinton's schedule as first lady.[52] Released by the National Archives as a non-searchable PDF, it covered eight years and totaled 17,481 pages.[53] The newspaper used the EC2 to convert the PDF to a searchable text—all within the same news cycle, for less than $150.[54] Second, in 2009, Pulitzer Prize-winning reporter Charles Duhigg began using the EC2 to store the 200 million data points he used in his "Toxic Waters" series.[55] His investigation explored the failures of the Clean Water Act and the Safe Drinking Water Act, and the EC2 allowed the paper to build a computerized database to guide Duhigg's reporting.[56] Third, in August 2013, the *New York Times* announced it had outsourced its email hosting to Google, moving all its reporters and editors to corporate Gmail accounts (previously, *Times* emails were stored on servers the newspaper owned).[57] Unlike the free Gmail service used by millions of consumers, corporate Gmail accounts cost money, and they offer greater privacy protections to their users, in a technological sense.[58] But the protections are not complete, and the move could leave *Times* reporters and their sources with fewer protections if they were targets of a government investigation.[59]

Widening the lens, then, the problem is that journalists and news organizations can benefit from using the cloud and other emerging technologies, but the government is enhancing its surveillance of electronic communication, which means, as noted earlier, whistleblowers and journalists committed to divulging secrets increasingly must depend on technologies and practices designed to keep secrets.⁶⁰ Notably, no form of technology is fundamentally private or public. Lawrence Lessig, the Roy L. Furman Professor of Law at Harvard Law School, was the first to articulate fully the idea that the Internet could be both liberating and constraining.

In his 2000 book *Code: And Other Laws of Cyberspace*, Lessig said software code and technical standards were a new form of law, because they shape what people can do using a particular system or network.⁶¹ This cut against the grain of the common belief that cyberspace is by nature unregulatable.⁶² Lessig said cyberspace has no nature—it has code, the software and hardware that make up the architecture of the Internet and Web.⁶³ Lessig argued that it is possible to choose what kind of cyberspace we want and what freedoms it guarantees.⁶⁴ Those choices are all about architecture: the code that governs cyberspace and who controls the code.⁶⁵ In that sense, Lessig said, code is "the most significant form of law, and it is up to lawyers, policymakers, and especially citizens to decide what values that code embodies," including the values of transparency, freedom of information, and privacy.⁶⁶

Second, Jack Balkin, the Knight Professor of Constitutional Law and the First Amendment at Yale Law School, used his 2013 article "The First Amendment is an Information Policy" to set out his views on the interaction of free speech theory, information policy, and the digital age.⁶⁷ He wrote that free expression exists in the larger context of policies for the spread of knowledge and information. "The practical ability to speak," Balkin wrote, "rests on an infrastructure of free expression that involves a wide range of institutions, statutory frameworks, programs, technologies, and practices."⁶⁸ He said we must "design democratic values into the infrastructure of free expression if we want an infrastructure that protects democracy."⁶⁹

A meta-summary of those works, along with the works of Rebecca MacKinnon, Tim Wu, Jeffrey Rosen, Evgeny Morozov, who all have written eloquently about the power and limitations of technology, would bear out many variations on the theme of regulating technology in a way that respects individual freedoms and the spirit of democracy. But conceiving and implementing a governance structure for that kind of digital world is a challenge. It is beyond the scope of this chapter to explore such a structure in its entirety, but one of its planks involves the values of transparency, freedom of information, and privacy as they apply to journalists. To the extent robust press coverage of government activities is a necessary ingredient of democracy, it is untenable for leak prosecutions and sweeping electronic surveillance to create or otherwise contribute to a "fearful atmosphere" with the proven ability to chill "contacts between American journalists and government sources."⁷⁰ That kind of atmosphere tips the balance significantly in favor of the government and stands to asphyxiate a large amount of substantive, independent reporting.

A hallmark of quality public-affairs journalism is the protection of confidential sources where the use of such sources is justified, so it is essential for journalists and news organizations to develop top-flight operational security skills (e.g., using anonymity technologies or using encryption to protect files as they are passed around). Doing so would also help the journalists and their organizations maintain relevance in a news and information ecosystem that includes the likes of WikiLeaks, OpenLeaks, TradeLeaks, Indoleaks, and Balkan Leaks—all highly secure and committed to publishing classified information from anonymous sources. Greenberg, the *Forbes* tech reporter, put it this way: "WikiLeaks brought to light a new form of whistleblowing that uses sophisticated cryptographic code to hide leakers' identities as they spill the private files of government agencies and coorporations."[71] If the mainstream press fails to take its security obligations seriously, it might be safer for leakers to take their business to WikiLeaks or one of its spin-offs.

LEGAL PROTECTIONS

Much more could be said about operational security (whether it should be taught in journalism programs, what news organizations have been doing, if anything, to train their journalists, and so on), but for now let's return to the idea of regulating technology to respect individual freedoms and the spirit of democracy. As noted earlier, nearly everything you do to communicate electronically leaves a trace.[72] Emails are stored on servers after they are deleted.[73] Phone calls create logs detailing which numbers connected, when they did so, and for how long.[74] Cellphones can create a record of where you are.[75] With that in mind, recall that the *New York Times* recently outsourced its email to Google.

On the one hand, a large email service provider like Google might offer better security than that built into a news outlet's internal server or network. On the other hand, when it comes to mounting a legal defense against a leak investigation, the *Times* could be making itself vulnerable. There will be a gap between the outlet and its data, and that gap could have serious implications. Before the *Times* outsourced its email to Google, generally the government would have been required to serve a subpoena on the *Times* in order to obtain its employee emails. Now a government investigator could try to serve Google, and if the subpoena came with a gag order, the *Times* might not learn of it timely or at all. "I worry a lot about the outsourcing of email at a news organization," *Wall Street Journal* reporter Julia Angwin told NPR, in August 2013. "We only have two layers of protection, right? One is technological, and one is legal. Certainly our lawyers…are gonna fight to protect our emails. But if they don't fully control them technically, they can't mount a very good argument."[76]

Putting aside the possible development of a federal shield law, the legal protections for journalists who use the cloud are, well, cloudy. There is a patchwork of doctrines and statutes that theoretically applies to the cloud, with each doctrine and statute applying in different circumstances. Among them are the Fourth Amendment,[77] the Stored Communications Act,[78] and the Electronic Communications Privacy Act.[79] They are discussed briefly below, followed by a deeper analysis of the Privacy Protection Act.

FOURTH AMENDMENT

The protections afforded by the Fourth Amendment, namely its warrant requirement, have not been extended fully to the cloud. The Fourth Amendment provides that the people shall "be secure in their persons, houses, papers, and effects, against unreasonable searches and seizures."[80] It also states that warrants must be issued "upon probable cause, supported by Oath or affirmation, and particularly describing the place to be searched, and the persons or things to be seized."[81] Requiring law enforcement to justify itself before conducting invasive searches offers a layer of constitutional protection that, if breached, renders the unlawfully seized evidence inadmissible in court against the person whose Fourth Amendment rights were violated.[82]

However, a warrant is not always necessary. A search can be executed without a warrant when items are in plain view, when a person consents to being searched, and so on. These are exceptions, and courts have developed a standard for determining when a search requires a warrant. It requires one only if there exists a "reasonable expectation of privacy" under the circumstances.[83] A person must have a subjectively reasonable expectation that an item was private, and the item must be something that society is willing to recognize as reasonably private.

Nuances have developed since that standard was articulated, and one is the "third-party doctrine."[84] It governs the collection of evidence from third parties in criminal investigations, and it says that by disclosing information to a third party, the subject gives up her Fourth Amendment rights in that information.[85] In other words, a person cannot have a reasonable expectation of privacy in information disclosed to a third party.[86] The rationale is that when you communicate with another person, you assume the risk that the other person will reveal the information to the public or law enforcement.[87] Thus, the third-party doctrine narrows the circumstances in which a warrant is required.

The key issue for extending Fourth Amendment protection to the cloud: Whether the cloud is something that society is willing to recognize as reasonably private. Societal expectations change with time, as technology and our uses of it change. With massive increases in bandwidth, wireless access, and mobile-device use in the past decade, remote storage has changed how the Internet is used.[88] Rather than being a purely public medium, the Internet has become a means of private storage and mobile or remote access. Changes in Internet usage suggest that society might be prepared to recognize a reasonable expectation of privacy in the cloud, at least in some circumstances.[89]

However, the third-party doctrine is a potential problem. When a journalist emails a source, is Google a party to that communication with regard to the email's content? Is the provider a party to a journalist's Google Docs spreadsheet with regard to the spreadsheet's content? If so, the government may be able to obtain that information without a warrant and without the journalist's consent. This issue is complicated by the fact that Google's advertising algorithms scan a user's emails for keywords in order to target advertising.[90] Does that make Google a party to the email with regard to the email's content? And a cloud service provider may reserve certain rights to access the contents of your account, according to the company's terms of service.[91] Does that make the provider a party to the contents? These questions remain unanswered.

STORED COMMUNICATIONS ACT

As courts have narrowed Fourth Amendment protections, Congress has passed legislation to fortify those protections and fill gaps created by new technologies. Enter: the Stored Communications Act (SCA), passed in 1986.[92] It protects wire and electronic communications stored by providers of an electronic communication service (ECS) or a remote computing service (RCS).[93] An ECS provides the ability to send and receive wire or electronic communications, and an RCS provides computer storage or processing services through wire, radio, electromagnetic, photo-optical or photo-electronic facilities. If a communication passes through a provider that does not qualify as an ECS or RCS, then the communication is not protected by the SCA.

The scope of the government's right to compel disclosure of a stored communication depends on (1) whether the provider is acting as an ECS or RCS, (2) whether the communication constitutes content or customer information, and (3) with regard only to an ECS, the number of days the communication has been in electronic storage.[94] For the latter, the government must obtain a warrant to obtain a communication stored by an ECS for not more than 180 days. If it has been stored for more than 180 days, the government may use a warrant or an administrative subpoena or court order.[95]

The SCA is convoluted, and its deficiencies have become apparent with the development of new technology. For example, having a lower standard of protection for emails stored longer than 180 days made sense in 1986 (back then email was available only through proprietary networks, using email was costly because the services charged based on connect time, and storage capacity was limited), but today, with the availability of free Web-based email accounts with nearly unlimited storage capacity, people routinely store emails longer than 180 days.[96] Moreover, protection for unopened emails is up for debate because of inconsistent interpretations of the term "storage" in the definition of an RCS.

In May 2011, Senator Patrick Leahy introduced the ECPA Amendments Act, which, among other things, proposed: (1) replacing the so-called "180-day rule" with a requirement that the government obtain a warrant for all stored communications, and (2) adding "geolocation information services" as a third category of regulated entities to address concerns raised by privacy groups regarding the protection of information available through mobile and GPS devices.[97] When he introduced the amendments, Leahy characterized the SCA as "significantly outdated and out-paced by rapid changes in technology" and that updating the law "is essential to ensuring that our federal privacy laws keep pace with new technologies and the new threats to our security."[98] Although the bill died when the 112th Congress ended, Leahy reintroduced it in March 2013.

ELECTRONIC COMMUNICATIONS PRIVACY ACT

The ECPA (it includes the SCA) consists of three parts. The first one, Title III, is the focus of this section. It prohibits the unauthorized interception of wire, oral, or electronic communications.[99] It also establishes a judicial procedure to permit such interceptions for law-enforcement purposes (this allows the government to intercept information about who you are, where you go, and what you do, all from data collected by cellphone providers, search engines, and social networks). The ECPA states that it is a federal crime to engage in wiretapping or electronic eavesdropping; to possess wiretapping or electronic eavesdropping equipment; to use or disclose information obtained through illegal wiretapping or electronic eavesdropping; or to disclose information secured through court-ordered wiretapping or electronic eavesdropping, in order to obstruct justice.[100]

The ECPA was a "forward-looking statute" when it was enacted in 1986. It afforded important privacy protections to subscribers of emerging wireless and Internet technologies.[101] The problem is that technology has advanced dramatically since 1986, and the ECPA has been left behind. The statute has not undergone a significant revision since the 1980s. As a result, the ECPA is a "patchwork of confusing standards that have been interpreted inconsistently by the courts, creating uncertainty for both service providers and law enforcement agencies."[102] The statute can no longer be applied in a clear and consistent way, and the vast amount of personal information generated by today's digital devices and services are no longer adequately protected.

PRIVACY PROTECTION ACT

For more than 30 years, the Privacy Protection Act (PPA) has enabled journalists in the offline environment to protect their work product. But would or should the PPA protect a journalist's cloud-based data from compelled disclosure? It's unclear whether the things journalists store in the cloud enjoy the same protection as the things they store on personal computers, on local servers, and in desk drawers.[103] By expanding the protections of the ECPA and SCA, the reforms would benefit journalists incidentally.[104] They would not, however, afford journalists the kind of protection they have come to expect from the PPA, which protects journalists from newsroom, computer, and other searches.[105] It applies to all kinds of law enforcement, and it prohibits the use of a search warrant to obtain materials from people engaged in First Amendment activities. Instead, the PPA requires the government to get a subpoena, giving the journalist a chance to challenge it in advance.[106]

The PPA states that with few exceptions "it shall be unlawful for a government officer or employee, in connection with the investigation or prosecution of a criminal offense, to search for or seize...any work product materials possessed by a person reasonably believed to have a purpose to disseminate to the public a newspaper, book, broadcast, or other similar form of public communication," or any "documentary materials possessed by a person in connection with" such a purpose.[107] Thus, the PPA divides materials into two categories: "work product" and "documentary." The former includes materials "prepared, produced, authored, or created" for dissemination to the public, like story drafts and outtakes.[108] The latter includes "materials upon which information is recorded," like photographs and videotapes.[109]

The PPA was passed in 1980, when the Internet consisted of a few thousand computers—14 years before Amazon was founded, 18 years before Google.[110] It was last amended in 1996, and it has not entered the cloud-computing era.[111] In fact, it is not clear the PPA has entered the Internet era. The statute has generated little case law, and it does not explicitly cover journalists at online-only outlets.[112] Many commentators have argued it does, citing the clause "other similar form of public communication."[113] But the only case to present that issue failed to resolve it: *Steve Jackson Games, Inc. v. United States Secret Service*, decided in 1994 by the Fifth Circuit.[114] The Secret Service searched the offices of a computer bulletin board operator and seized computers, disks, and other materials.[115] The court held that the agents violated the PPA but based its decision on the fact that the company also published books and magazines, explicitly covered by the PPA.[116] It would be sensible to amend the PPA to cover—explicitly—people who use the Internet to communicate with the public.[117]

The cloud raises other PPA questions, too. The act's protection is countenanced in terms of possession.[118] Who "possesses" the information in the cloud?

There's no case law directly on point, but, putting aside the third-party doctrine from the Fourth Amendment context, it seems logical that if a journalist stored a document in the cloud, the document would be hers. The cloud would act as her digital desk drawer, and the presence of the cloud provider—a mere intermediary, without control of the document's content—would not destroy the journalist's possession.[119] For these reasons, it would be sensible to amend the PPA to protect—in terms of possession—the data that journalists store in the cloud. But if Congress decides users do not "possess" their cloud-based data, potentially allowing the government to request the data from the provider, the PPA should be amended to require the government to give some form of notice so the journalists can challenge the subpoena—and sue for damages if the government fails to give notice.[120]

CONCLUSION

Alan Dershowitz, the Felix Frankfurter Professor of Law at Harvard Law School, wrote in his 2013 memoir that "no reasonable person can dispute the reality that there are necessary secrets" and that no student of history can "doubt that there are unnecessary secrets."[121] To him the real issue is not whether secrets should be exposed—it is who should be entrusted to make the decision to expose them.[122] Add to that the issue of who is in a position to make that decision, whether or not she is entrusted to do so. Traditionally, a leaker or whistleblower worked at the government's highest levels and provided information to a reporter at some venerable news organization: the *New York Times*, the *Washington Post*, the *Chicago Tribune*. Relatively few people were in a position to leak, and relatively few people were in a position to publish a leak. Today, the field is far more open.

The state of the world's information favors the leaker more than ever, because digitally recorded data make up the vast majority of the world's recorded information. Those data are reproducible and mobile, and in terms of classified information, it is no longer necessary to be a war hero who reaches the government's highest levels in order to get access to the data. Rather, a leaker could be one of the millions of Americans who has access to secret government information. And for secrets it is a seller's market.

In addition to the venerable news organizations that for decades have worked with leakers, there exists today a variety of people and entities in a position to publish a leak. Digital media have democratized publishing, and WikiLeaks pioneered a new form of whistleblowing that uses anonymity technology to hide leakers' identities as they spill. The site inspired a number of spin-offs, including OpenLeaks and TradeLeaks, and they all are committed to publishing secrets from anonymous sources. The social power of technology is reshaping the gathering, production, and distribution of secrets.

At the same time, in the United States aggressive prosecution of leakers and broad electronic surveillance and data-mining programs are detering government officials from speaking to journalists. The officials believe the investigations and surveillance have made it more difficult for journalists to protect them as sources. It is a harsh new reporting environment that demands operational security skills and legal protections—regulating technology to respect individual freedoms and the spirit of democracy.

Putting aside the possible development of a federal shield law, the legal protections for journalists who use the cloud are unsettled. Various doctrines and statutes theoretically apply to the cloud, which, for security reasons, might never be the best place for journalists and others to store highly sensitive information. The SCA, ECPA, and PPA are badly in need of reform. And, more generally, it is time for President Obama to fulfill his very first promise—to make his administration the most transparent in American history, in the spirit of transparency for the sake of accountability.[123] Whether he does so will have a lasting impact on government accountability and the standing of the United States as a global beacon of press freedom.[124]

NOTES

1. Ravi Somaiya, "Editor Describes Pressure After Leaks by Snowden," *New York Times*, Dec. 3, 2013, available at http://www.nytimes.com/2013/12/04/business/media/after-snowden-revelations-a-changed-world-for-journalists.html
2. *Id.*
3. *Id.*
4. *Id.*
5. *Id.*
6. *Id.*
7. *Id.*
8. *Id.*
9. *Id.*
10. Andy Greenberg, *This Machine Kills Secrets* 6 (Dutton 2012).
11. *Id.* at 5.
12. *Id.*
13. *Id.*
14. *Id.*
15. *Id.*
16. *Id.*
17. *Id.*
18. *Id.* at 46.
19. *Id.*
20. *Id.*
21. *Id.* at 5.

22. *Id.*
23. *Transparency and Open Government, Memorandum for the Heads of Executive Departments and Agencies*, The White House, Jan. 21, 2009, available at http://www.whitehouse.gov/the_press_office/TransparencyandOpenGovernment
24. Jonathan Easley, "Obama says his is 'most transparent administration' ever", *The Hill*, Feb. 14, 2013, available at http://thehill.com/blogs/blog-briefing-room/news/283335-obama-this-is-the-most-transparent-administration-in-history
25. Greenberg, *supra* note 10, at 225.
26. *Id.* at 5.
27. Kira Goldenberg, "Obama's broken promises on transparency," *Columbia Journalism Review*, Oct. 10, 2013, available at http://www.cjr.org/behind_the_news/cjp_report_on_us_press_freedom.php
28. Greenberg, *supra* note 10, at 224.
29. *Id.*
30. *Id.*
31. *Id.*
32. Cora Currier, "Charting Obama's Crackdown on National Security Leaks", *ProPublica*, July 30, 2013, available at http://www.propublica.org/special/sealing-loose-lips-charting-obamas-crackdown-on-national-security-leaks
33. *Id.*
34. *Id.*
35. Jonathan Peters and Edson C. Tandoc, Jr., "'People Who Aren't Really Reporters at All, Who Have No Professional Qualifications': Defining a Journalist and Deciding Who May Claim The Privileges," *N.Y.U. Journal of Legislation and Public Policy-Quorum*, October 8, 2013, 34, 35.
36. *Id.*
37. *Id.* at 34–35.
38. Currier, *supra* note 32.
39. *Id.*
40. Leonard Downie, Jr., *The Obama Administration and the Press*, Special Report, Committee to Protect Journalists, Oct. 10, 2013, available at http://www.cpj.org/reports/2013/10/obama-and-the-press-us-leaks-surveillance-post-911.php
41. *Id.*
42. *Id.*
43. *Id.*
44. James Bamford, "The NSA Is Building the Country's Biggest Spy Center (Watch What You Say)," *Wired*, March 15, 2012, available at http://www.wired.com/threatlevel/2012/03/ff_nsadatacenter/
45. *Id.*
46. *Id.*
47. *Id.*
48. *Id.*
49. *Id.*
50. *Id.*
51. Jonathan Peters, "Updating the Privacy Protection Act for the Digital Era," *Columbia Journalism Review*, Jan. 30, 2012, available at http://www.cjr.org/behind_the_news/updating_the_privacy_protectio.php?page=all
52. *Id.*
53. *Id.*

54. *Id.*
55. *Id.*
56. *Id.*
57. Steve Henn, "Switching To Gmail May Leave Reporters' Sources At Risk," *NPR*, August 16, 2013.
58. *Id.*
59. *Id.*
60. Greenberg, *supra* note 10, at 6.
61. Lawrence Lessig, *Code: And Other Laws of Cyberspace* (Basic Books 2006).
62. *Id.*
63. *Id.*
64. *Id.*
65. *Id.*
66. *Id.*
67. Jack Balkin, "The First Amendment Is an Information Policy," *Hofstra Law Review* 41 (February 2013).
68. *Id.*
69. *Id.*
70. Downie, *supra* note 40.
71. Greenberg, *supra* note 10.
72. Henn, *supra* note 57.
73. *Id.*
74. *Id.*
75. *Id.*
76. *Id.*
77. Fourth Amendment to the U.S. Constitution.
78. 18 U.S.C. 121 §§2701–2712.
79. 18 U.S.C. §§2510–2522.
80. Fourth Amendment to the U.S. Constitution.
81. *Id.*
82. David A. Couillard, "The Cloud and the Future of the Fourth Amendment," *ArsTechnica*, April 26, 2010, available at http://arstechnica.com/tech-policy/2010/04/the-cloud-and-the-future-of-the-fourth-amendment/
83. *Id.*
84. *Id.*
85. *Id.*
86. *Id.*
87. *Id.*
88. *Id.*
89. *Id.*
90. *Id.*
91. *Id.*
92. 18 U.S.C. 121 §§2701–2712.
93. *Id.*
94. *Id.*
95. 18 U.S.C. 121 §§2701–2712.

96. *Id.*
97. *Id.*
98. *Leahy Introduces Benchmark Bill To Update Key Digital Privacy Law*, Office of U.S. Senator Patrick Leahy, May 17, 2011, available at http://www.leahy.senate.gov/press/leahy-introduces-benchmark-bill-to-update-key-digital-privacy-law
99. 18 U.S.C. §§2510–2522.
100. *Id.*
101. *ECPA Reform, About the Issue*, Digital Due Process, 2010.
102. *Id.*
103. Peters, *supra* note 51.
104. *Id.*
105. *Id.*
106. *Id.*
107. *Id.*
108. *Id.*
109. *Id.*
110. *Id.*
111. *Id.*
112. *Id.*
113. *Id.*
114. *Steve Jackson Games, Inc. v. United States Secret Service*, 816 F.Supp. 432 (W.D.Tex., 1993).
115. Peters, *supra* note 51.
116. *Id.*
117. *Id.*
118. *Id.*
119. *Id.*
120. *Id.*
121. Alan Dershowitz, *Taking the Stand: My Life in the Law* (NY: Crown Publishing, 2013): 142.
122. *Id.*
123. Downie, *supra* note 40.
124. *Id.*

CHAPTER NINE

Finding Resolution: Systems for Resolving Disputes and Reconciling Access with Privacy

DAXTON R. "CHIP" STEWART

Conflict is inevitable in human affairs, and it is particularly present when parties with competing interests interact with divergent notions of what is right. Unsurprisingly, the interests at stake in public records and meetings—transparency on the one hand, privacy on the other—compete against one another in challenging, and sometimes destructive, ways.

These competing areas—privacy and transparency—are hallmarks of American democracy. Both have emerged as core human rights over the past century. The right to privacy, famously recognized more than a century ago as "the right to be let alone,"[1] is not explicitly outlined in the Bill of Rights, but it has become recognized as a right through a series of Supreme Court decisions,[2] supplemented by statutes such as the federal Family Education Rights and Privacy Act and the Healthcare Information Portability and Accountability Act. The right to know, of course, has also become recognized as a policy central to good democratic practice with constitutional roots, "an integral part of the system of freedom of expression, embodied in the First Amendment and entitled to support by legislation or other affirmative government action."[3] While not explicitly guarded by the Bill of Rights, the right of access to public records and meetings is enshrined in Article I, Section 24 of the Florida Constitution and is guarded by public records and open meetings laws in every state and at the federal level.

Our understanding of these issues as essential rights—the right to know and the right to privacy—grants parties that seek to protect and assert them great power, with true belief that what they are guarding is central to core American values. But the high value parties place on these rights also heightens their conflict potential. Opposing camps fall into classic notions of conflict, the "perceived divergence of interest, a belief that parties' aspirations are incompatible."[4] If competing parties cannot be moved from their positions, it can lead, and has very clearly led, to persistent, intractable conflict between those who would seek more openness and access and those who would seek to protect informational and personal privacy. Further complicating this conflict is that the parties that argue on either side do not always include important stakeholders. News media and transparency advocates may argue for the right to know, when the primary beneficiaries of such information may be interested or affected citizens. Meanwhile, government employees may argue on behalf of protecting privacy, while those individuals whose privacy would be jeopardized are not part of the discussion.

The roots of this conflict play out in the daily disputes that arise between those seeking access to government information and those denying it on grounds of legal exemptions, protecting personal privacy, or other reasons. The various state and federal public records and open meetings laws discussed elsewhere in this book provide a framework for, as Hawaii's Uniform Information Practices Act puts it, balancing "the individual privacy interest and the public access interest, allowing access unless it would constitute a clearly unwarranted invasion of personal privacy."[5]

These legal frameworks also provide ways of resolving disputes that arise, though in several very different fashions. All of the public access statutes allow parties denied access to seek a remedy in the courts, but the federal government and nearly two-thirds of the states also provide alternative methods of dispute resolution, either through enhanced powers by the attorney general, special bodies charged with monitoring public access disputes, ombuds offices, informal mediation programs, or combinations of these programs. Such programs appear to be growing in popularity. In recent years, Pennsylvania has implemented an Office of Open Records that has powers to mediate disputes and to issue advisory opinions; Maine has created a Public Access Ombudsman to resolve complaints and issue advisory opinions; and the federal government has opened the Office of Government Information Services in the National Archives and Records Administration to oversee federal agency compliance with the Freedom of Information Act and to resolve disputes between requesters and agencies.

Such extra-judicial systems are representative of the alternative dispute resolution movement, which has developed over the past quarter century, partially in response to the cost, delay, and other weaknesses inherent in litigation. Conflict resolution and systems design scholars and professionals have applied their understanding of how to improve negotiation, manage and process disputes, and even transform conflict through such non-judicial processes, with some scholarship over the past decade examining disputes arising under public access laws.

The purpose of this chapter is to examine some of the public access dispute resolution systems and how they handle the balance between transparency interests and privacy interests. It begins by exploring the spectrum of dispute resolution options available, and then moves on to propose other new ways of thinking about managing conflict about transparency and privacy at both the local and online context.

SPECTRUM OF DISPUTE RESOLUTION OPTIONS

One way to think of the different ways to resolve disputes is as a spectrum, ranging from less formal to more formal options. On the "less formal" end would be informal discussion and negotiation, which carry very little cost or time and also have almost no authority to bind disputing parties to their outcome. On the other end would be litigation, with a greater level of formality, high costs and time commitments to the parties, and more power to result in binding decisions.

INFORMAL <--> FORMAL

Negotiation----Mediation----Ombuds----Administrative Adjudication----Litigation

These extremes, and any number of points in between, are represented in the range of options states and the federal government offer for handling disputes that arise under public access laws. Dispute systems design theorists suggest that ideal systems include a range of options, moving from less formal to more formal, with parties being able to find a place to resolve their disputes with the least amount of time and cost commitment necessary.[6] Some states—Connecticut and Pennsylvania, for example—provide for such a range of options. Beginning with the informal, below are some of the options states currently employ.

NEGOTIATION

Informal negotiation is always available as an option, and is often the first and most reasonable option for people seeking access to public records. Negotiation can begin even before the process becomes formalized, such as when a citizen makes an oral request for access to a record before submitting a written request as outlined by most open government laws.

In most places, particularly those without more formalized dispute resolution options outside of litigation, informal negotiation systems have emerged among advocacy and interest groups. For example, in Georgia, citizens may go to the Georgia First Amendment Foundation to seek the help of a well-connected director who can make calls or otherwise lean on public officials before those who have been denied access seek more formal procedures through the attorney general or the courts. Similarly, in Missouri, journalists may seek the help of the attorney for the Missouri Press Association, who can help with access negotiations.[7]

Similarly, some of the informal systems are tied to government institutions, without mandates built into public access laws. These include Open Meetings Law Enforcement Team (OMLET), established in the Arizona attorney general's office, which works to resolve disputes over access to meetings of government bodies. Washington established an Open Government Ombudsman in the attorney general's office in 2005 to offer support to citizens and government agencies by responding to inquiries with advisory opinions and making calls to help resolve disputes. Even at the federal level, informal negotiation can be initiated by the Office of Government Information Services, a department established in 2007 under the National Archives and Records Administration to help requesters navigate the Freedom of Information Act.

However, the lack of formality of these processes leaves much to be desired. The process is entirely voluntary, so if any party involved chooses to opt out of the negotiation, there is no recourse besides a more formal option. In some jurisdictions, mediation programs have been authorized by statute to help disputing parties resolve their differences.

MEDIATION

Mediation typically involves a third-party neutral who works with disputing parties through a voluntary, non-binding process to seek common ground and reach agreements that maximize the interests of the parties.[8] Mediation has become one of the most popular forms of alternative dispute resolution, with programs developed in state and federal courts, and it is mandated under the Alternative Dispute Resolution Act for use in federal regulatory agencies. Its supporters assert that mediation is less time consuming, more cost-efficient, and more satisfactory to disputing parties.[9]

While, technically, mediation is informal in nature—any set of disputing parties can voluntarily agree to mediate any dispute without resorting to judicial processes—mediation has become a bit more formalized through their inclusion in various courts and regulatory programs, including public access and privacy disputes.

Florida was a pioneer in establishing its public records mediation program, which began as an informal program in the attorney general's office and later became more formalized when it was authorized by statute. Under the program, the attorney general or another party can serve as a mediator between government officials and people seeking access to records when those parties voluntarily agree to participate. In the most recent numbers provided, the program announced a 75 percent success rate, reaching a non-binding agreement in 60 out of 80 cases mediated in 2007.[10]

Other states have initiated informal mediation programs as well, including Pennsylvania, which offers mediation of disputes through the Office of Open Records[11] during the appeals process offered by government agencies that have initially denied access.

Another hallmark of mediation is confidentiality of the process, which is aimed at allowing parties to participate openly and in good faith without fear of outside consequences for their efforts at reaching settlement. Mediators are typically bound to keep the discussions and results of the process private. This presents a challenge for disputes arising under public access laws because they involve government employees, whose conduct is usually deemed open to public access under these laws.

In Pennsylvania, parties cannot make the result of the open records mediation private. This has the benefit of not allowing parties to cloak their agreements in secrecy. However, because mediation is a voluntary process, it may also work to discourage parties, particularly government bodies, from using the program. If the outcome of the appeal will become public regardless of the method used to resolve the dispute, a public agency may opt for a lengthier process or one with a better chance of success than a non-binding, voluntary process such as mediation.

Mediation of open government disputes does not mirror more traditional mediation in other fields. Rather than using an independent mediator, open government mediations are often conducted by officials in the attorney general's office or the public access-specific administrative body. And open government mediations are more typically done via phone than face-to-face. Nevertheless, they are inexpensive programs to implement, generally with no cost to the disputing parties, and can work to divert disputes from more formal, time-consuming and expensive processes such as litigation.

OMBUDS

Classically, an ombuds is an independent government official in charge of providing oversight of government bodies, investigating conduct, and issuing recommendations.[12] A number of states have implemented ombuds offices to oversee compliance with public access laws.

However, not all public access ombuds have the same powers and duties. Some are perhaps best thought of as quasi-ombuds offices, which have the power to hear complaints, resolve disputes, and issue advisory opinions and other recommendations, but do not have formal powers to investigate alleged misconduct or to issue binding decisions. Examples of the quasi-ombuds model are New York's Committee on Open Government, Virginia's Freedom of Information Advisory Council, and the Public Access Counselor's offices in Illinois and Indiana, which are sometimes referred to as "ombuds" programs. The flexibility and relatively low cost of maintaining quasi-ombuds offices such as these have made them an option appreciated both by government agencies and openness advocates.[13]

More traditional ombuds offices have been established in the ombudsman or citizens' aide offices in Iowa and Arizona, which authorized special positions in those offices to serve as the primary overseer of open government activities and disputes. However, while these offices have more investigative power and had a bit more difficulty than the quasi-ombuds programs, in part because just one person has been committed to handle all of the state's public access disputes, and perhaps also because even with investigative authority, neither has the ability to make final decisions or recommendations that are binding on the parties. Iowa, for example, after years of having one ombuds office employee oversee public records and open meetings issues, established a Public Information Board in 2012 that would have power to prosecute contested cases arising under state open government laws, while also performing informal mediation and creating training programs.[14]

In 2009, the most ambitious public access ombuds project was launched when the federal Office of Government Information Services began operations. An amendment to the Freedom of Information Act created the office, which is charged with reviewing FOIA policies and procedures of agencies, reviewing agency compliance with FOIA, and mediating disputes as an alternative to litigation.[15] While director Miriam Nisbet[16] has said that she considers the office's primary role to be a mediator, the additional powers and accomplishments of the office—such as creating an online dispute tracking system for people to follow their FOIA requests—go beyond the role of the traditional mediator.

Ombuds offices are a mix of the formal and informal. Ombuds can engage in telephone diplomacy to informally mediate disputes upon calls from requesters or government employees, or they can issue more formal but non-binding advisory opinions, and some can conduct audits and investigations into agency compliance. This range of formal and informal roles makes the ombuds a good fit for many jurisdictions, though several others have adopted more formalized appeals processes through other public access-specific bodies.

ADMINISTRATIVE ADJUDICATION

Several states have established access-specific agencies such as public access counselors, open government boards, and procedures through the attorney general's office to handle government transparency and other records management issues.

These include programs that are "litigation lite," with fact-finding and appeals processes that are more binding in nature and have developed their own bodies of precedent.

The longest tenured of such offices is the Connecticut Freedom of Information Commission, established in 1975. The commission has powers that are ombuds-like in nature, with the ability to investigate and mediate disputes arising under the state public access laws. However, the commission is also authorized to operate an appeals process, which must be completed by disputing parties before they can seek judicial remedies in court.[17] Connecticut's FOI Commission is a model representation of the dispute systems design ideal of implementing mixed methods ranging from less formal to more formal options along the dispute resolution spectrum.

A similar formalized appeals process has been established in Utah, overseen by the State Records Committee, which is like Connecticut's program but without informal mediation options. Utah's program has the power to make binding decisions, and can hold expedited hearings and appeals in a matter of weeks rather than the months it would take in court.[18] However, the process has not been used very often. The committee has typically issued 15 to 20 final decisions each year since 2007.[19]

In Texas, the administrative appeal is run through the attorney general's office. If a government body wants to deny access to a requester under the Public Information Act on any potentially disputable matter, it must seek a letter opinion from the attorney general authorizing the denial.[20] The attorney general issued more than 19,000 such letter rulings in both 2010 and 2011.[21]

All of the aforementioned options—negotiation, informal or formal mediation, intervention by an ombuds or quasi-ombuds office, or appeals through public access-specific administrative bodies—are offered in addition to litigation, the most traditional option of resolving disputes.

Overall, nearly two-thirds of the states offer some form of alternative to litigation,[22] but litigation should not be dismissed as an unreasonable option. While they may be more time-consuming and costly, judicial processes are necessary to help shape the law through appellate court decisions and to preserve the rights of parties who would abuse the judicial process or extra-judicial processes as a way of delaying or frustrating opponents. The alternative dispute resolution field is constantly concerned that its non-judicial options overly favor resolution and settlement, perhaps at the expense of protecting legal rights and building a consistent body of law.[23]

But both systems rely upon one another. The judicial process can use alternative dispute resolution options to relieve stress on overly burdened dockets, while non-judicial processes sometimes cannot reach voluntary agreements and need an independent decision-maker with the authority to issue a binding ruling.

LOOKING FORWARD

The processes mentioned above have had varying degrees of success, but there are other ways of thinking about resolving conflict between right-to-know advocates and privacy advocates. Below are two that could be considered—first about addressing conflict at the micro level, and, second, by sidestepping conflict altogether through technological systems.

Negotiated Local Protocols

Moving parties away from escalated conflict and stalemate is challenging, particularly as conflict has lingered and the parties persist in destructive cycles of conduct and mutual distrust. Resolving this kind of conflict takes more than legislation or national movements by advocacy groups. It takes individuals at the local level—those stakeholders who have been most involved in the persistent conflict—finding ways to move past old disputes and contentious, competitive behaviors and into more productive, cooperative patterns.

There is no better way to do this than by getting disputing parties face-to-face, where old disputes and grudges can be aired and a path forward can be developed. A number of years ago, this author contacted a number of government officials and openness advocates in a small Midwestern town who had long battled over the balance between transparency and privacy, what the public had a right to know and what the government had no duty to provide.

The contacts included the records custodians of the local police department, county sheriff's department, the university police department, and area print and broadcast news media outlets. Each expressed dismay at the way records requests were made and handled, while also admitting some confusion as to what the law required. Each also expressed a willingness to get together at a roundtable to talk about their differences, their interests, their priorities, and ways that the records request and processing system could be made more efficient and satisfactory.[24]

Such a willingness to get together, to work on understanding the interests of other parties, and to work on developing a local protocol for processing records requests is a strong first step that can be undertaken in any community. Public policy dispute resolution procedures, such as negotiated rulemaking, have been "enormously successful" when employed by federal agencies "in developing agreements in highly polarized situations and has enabled the parties to address the best, most effective, or most efficient way of solving a regulatory controversy."[25] Of course, the present situation is not exactly a "regulatory controversy." Instead, it regards application and enforcement of a law that has proven to be extremely difficult to manage efficiently through judicial processes. Further, local protocols have the advantage of understanding local culture and relationships in a way that state and federal laws do not.

While all negotiated rulemaking would be done in light of the relevant open government and privacy laws, participants in the rulemaking procedure could work to fit the protocols to the interests of the parties without having to trigger the laws. For example, consider a protocol on timely responses to routine open records requests from news media organizations. While a state may have a law outlining a procedure for a printed, written request, followed by up to five business days for the government agency to provide a written response and access to the records in question, this would obviously serve to frustrate journalists on deadline. A local protocol could be developed for release of police reports on the same day via phone or email request, with victims' names and other information triggering privacy concerns redacted. Denials could be issued same day as well to let requesters know if more formal processes would be needed to resolve disputes.

By including an independent convener, sometimes a trained mediator or facilitator, local groups could develop a body of rules agreed upon by the most important stakeholders—openness advocates, privacy advocates, news media, and government record-keepers—through a consensus-building process that could help divert future disputes from other, more formal processes.

Web-Based Open Government Portals

One challenge to transparency is in the way public access laws are structured. While the laws typically create a legal presumption of openness—that records and meetings are open unless the government can clearly establish an appropriate exemption—the process for allowing citizens access operates in exactly the opposite manner. In essence, the system for creating and maintaining records begins with closure. The records are kept out of the public eye by government bodies, which then must allow access to records via request through the process outlined in the appropriate public records statute. It is only after a proper request has been filed, and the records custodian has determined that the record does not fall under any exemption, that access is permitted. This approach cannot help but to invite conflict between those who seek access to information and those who manage access.

But what if this level of conflict could be avoided? Public access laws could be structured in a way that demanded that public records be made available publicly upon their creation, through a system that separates public and private documents at the outset, rather than waiting for a request and the dispute that follows. The web makes it possible for records to be posted immediately in a way that gives power to citizens to view and analyze them, rather than the current system of placing records on hard drives, or in hard copies behind desks, where clerks, custodians and other government employees serve as gatekeepers.

In almost every state, electronic release of requested public documents is recommended and may be done at the discretion of the records custodian.

However, New Mexico amended its public records act to mandate creation of a website that is "free, user-friendly, searchable, and accessible to the public."[26] New Mexico's Sunshine Portal opened in July 2011 and includes records such as the state's cash balance on hand, annual budgets for state agencies, government purchases and revenue, and employee salaries.[27] State agencies are required to provide updated information as frequently as possible, but no less than once each month.

The federal government has also been working on creation of an online portal, though of a different type. The Environmental Protection Agency, the Department of Commerce, and the National Archives and Records Administration were working in early 2012 on establishing "a multi-agency FOIA portal that automates FOIA processing and reporting, stores FOIA requests and responses in a repository, and keeps records electronically."[28] The site would allow citizens to track their own requests and to view agency responses online.

Currently, such portals trigger few personal privacy concerns. The information released on New Mexico's page is more informational in nature, with financial information such as budgets, income, and expenditures. The portal does not, for instance, include personnel files, police reports, medical examiner records, or other items that can trigger exemptions or redaction of portions of government records. More comprehensive government records portals may be able to include such information, but it would have to be overseen manually by a person trained to exempt or at least flag such information, or it would have to be carefully coded so such records could be processed more automatically.

In the future, it would be ideal for states and the federal government to develop electronic records management systems that do what New Mexico and the federal government are trying to accomplish. New Mexico's portal potentially reduces conflict through avoidance by making records available for analysis outside of the realm of requesters and records custodians. However, when formal requests must be made and processed, the federal government portal would provide tracking of them, potentially allowing more efficient handling of appeals and disputes. Further, the director of the Office of Government Information Services has suggested developing an online dispute resolution (ODR) system as part of its mediation efforts.[29] Online dispute resolution uses traditional ADR tools such as arbitration and mediation and develops electronic systems to allow them to proceed online, sometimes with computer assistance in organization and making negotiation and dispute processing more efficient. It can include automated systems to resolve disputes as well. ODR has been successfully used by companies such as PayPal and eBay and may be an option for resolving public access disputes when parties are separated by great geographic distances and desire prompt resolution.

This would require both financial investment in electronic records management software systems and in securing enough online storage space to facilitate access. It also would require increased uniformity in record-keeping, a challenge when state and local agencies use a variety of programs developed by private vendors to create and manage recordkeeping.

CONCLUSION

Fifty years from now, how will public access to records and meetings be balanced with personal and informational privacy rights? It is inevitable that technology will play a greater role in how government records are created and stored and how meetings are conducted. It is similarly hard not to imagine that the laws will struggle to adapt to the technology but must eventually embrace it to maximize the efficiency of government operations.

Looking forward, it is essential that transparency advocates work to develop flexible systems that maximize citizen access, reduce the barriers to access of routine government records and meetings through delay or misapplication of exemptions, and reduce the time it takes to appeal denials.

Equally essential is ensuring that such systems, in their move to maximize efficiency and access, continue to protect personal and information privacy in a way that people have come to expect. No technological system will be able to account for the complexities of people's privacy concerns, so it will take a combination of good software design, well-trained government records custodians, and a flexible, efficient dispute resolution system to achieve the balance.

NOTES

1. Samuel D. Warren and Louis D. Brandeis, "The Right to Privacy," *Harvard Law Review* 4 (December 1890): 193–220.
2. *Griswold v. Connecticut*, 381 U.S. 479 (1965); *Reporters Committee for Freedom of the Press v. Department of Justice*, 489 U.S. 749 (1989); *Lawrence v. Texas*, 539 U.S. 558 (2003).
3. Thomas I. Emerson, "Legal Foundations of the Right to Know," *Washington University Law Quarterly* 1 (1, 1976): 2.
4. Dean G. Pruitt, Jeffrey Rubin, and Sung Hee Kim, *Social Conflict: Escalation, Stalemate, and Settlement* (New York, NY: McGraw-Hill, 2003): 6–7.
5. Hawaii Revised Statutes, sec. 92F-2(5) (2011).
6. Cathy A. Costantino and Christina Sickles Merchant, *Designing Conflict Management Systems: A Guide to Creating Productive and Healthy Organizations* (San Francisco, CA: Jossey-Bass, 1996); William L. Ury, Jeanne M. Brett, and Stephen B. Goldberg, *Getting Disputes Resolved: Designing Systems to Cut the Costs of Conflict*, (San Francisco, CA: Jossey-Bass, 1988).
7. Daxton R. Stewart, "Systems in the Shadow of Sunshine Laws," *Rutgers Conflict Resolution Law Journal* 9 (Spring 2012), 1–38.
8. American Bar Association, "Model Standards of Conduct for Mediators," http://www.americanbar.org/content/dam/aba/migrated/dispute/documents/model_standards_conduct_april2007.authcheckdam.pdf

9. Roselle L. Wissler, "Court-Connected Mediation in General Civil Cases: What We Know from Empirical Research," *Ohio State Journal on Dispute Resolution* 17 (3, 2002): 641–703.
10. Commission on Open Government Reform, "Reforming Florida's Open Government Laws in the 21st Century," p. 62, http://www.floridafaf.org/resources/open-government-reports/commission-on-open-government
11. Pennsylvania Office of Open Records, "Informal Mediation," http://openrecords.state.pa.us/portal/server.pt/community/open_records/4434/informal_mediation/488137
12. Howard Gadlin, "Ombudsman: What's in a Name?" *Negotiation Journal* 16 (January 2000): 37–48.
13. Daxton R. Stewart, Systems in the Shadow of Sunshine Laws. *Rutgers Conflict Resolution Law Journal* 9 (2, 2012): 1–39.
14. Iowa Public Information Board Status Report (February 2013), https://governor.iowa.gov/wp-content/uploads/2012/07/IPIB-Annual-Report-2013.pdf
15. United States Code 5 (2011), sec. 552(h).
16. Miriam Nisbet, "Statement Before the United States Senate Committee on the Judiciary," (2009) https://ogis.archives.gov/Assets/Website+Assets/News+and+Events/09-09-30-nisbet.pdf
17. Connecticut Statutes, sec. 1-206(d) (2011).
18. Utah Code, sec. 63-2-403 (2011).
19. Utah State Records Committee, "State Records Committee Appeal Decision Summaries," http://archives.utah.gov/src/srcappeals-2010–2012.html
20. Texas Government Code, sec. 552 (2011).
21. Attorney General of Texas, "Open Records Letter Rulings," https://www.oag.state.tx.us/open/index_orl.php?ag=50abbott&fmt=htm
22. Daxton R. Stewart, "Managing Conflict over Access: A Typology of Sunshine Law Dispute Resolution Systems," *Journal of Media Law & Ethics* (winter-spring 2009): 49–82.
23. Wayne D. Brazil, "Rights and Resolution in Mediation: Our Responsibility to Debate the Reach of Our Responsibility," *Dispute Resolution Magazine* 16 (summer 2010): 9–12.
24. Chip Stewart, "Negotiated Rulemaking and the Sunshine Law: Can it help local law enforcement and the press get along?" unpublished manuscript (2006), http://works.bepress.com/daxton_stewart/10/
25. Philip J. Harter, "Assessing the Assessors: The Actual Performance of Negotiated Rulemaking," *New York University Environmental Law Journal* 9 (2000): 38.
26. New Mexico Statutes, sec. 10-16D-3 (2011).
27. New Mexico Sunshine Portal 2012, http://sunshineportalnm.com
28. Office of Government Information Services, "FOIA Portal Moving from Idea to Reality," http://blogs.archives.gov/foiablog/2012/01/09/foia-portal-moving-from-idea-to-reality/
29. Nisbet, "Statement Before the United States Senate Committee on the Judiciary."

CHAPTER TEN

Here's Looking at Me: The Abandonment of Privacy and Solitude as Millennials Move to Life Online

PAUL GATES

"You have zero privacy anyway. Get over it."
SCOTT MCNEALY, CEO, SUN MICROSYSTEMS[1]

An old legal aphorism holds that the law lags technologically driven social change by a generation, but that view was conceived in an era when the measure was a human generation. Today, generations refer more commonly to technological advances, and in many areas the law was left behind long ago. In fact the new lingua franca of the digital generation, transparency, has become the ultimate virtue; only Luddites, still mired in the last decade of the 20th Century cling to dated notions of privacy.

Young people have always been concerned with image, and for generations carefully cultivated personal identities through (among other cultural signifiers) clothing, hairstyles, and behaviors generally presented to a relatively small group of friends. Since the debut of Facebook, however, the reach of individual identities has increased exponentially. And Twitter's ability to let users "follow" celebrities causes a breakdown in social distance that leads in some cases to fans actually following—if not outright stalking—the objects of their obsession.

The ability to reveal personal information and images to "friends" and carry on relationships in numbers far beyond what would be possible to maintain in actual friendships is also fraught with peril. Fact, rumor, and photos in cyberspace live on long after memory has faded and the relationships have petered out. Technological documentation, recall, and 24/7 availability worldwide pose a threat to personal privacy and reputation far beyond even the openness inherent in the intimacy of the communal ties of pre-industrial village life.

It is beyond naïve to think that free access to the technological wizardry that allows us to spread comments weighty and otherwise is solely for entertainment, or that the ability to call up facts both useful and obscure is designed primarily to further education. Hardware sophisticated enough to process billions of bits of personal data and the software engineers talented enough to design systems to capture them are expensive. Once compiled, they become a valuable commodity that can be readily "monetized" through sales to commercial interests that can then push narrowly targeted advertising back through the same cyber-channels.

Constant connectivity—what sociologists term "ambient awareness"—has also destroyed the concept of solitude, as young people spend more and more time positioning their online presence and fine-tuning new angles to curate their brand for a larger and larger base, thereby solidifying their virtual identities.

This chapter examines the changing norms and expectations of a new generation of citizens, the Millennials. The battles in government over balancing government access with personal privacy will no doubt be affected by the mood of the public, which is changing by the day. As a technologically adept generation quickly masters the newest tools, uses and possibilities are immediately apparent. Unintended consequences may not occur to users as quickly as novelty and the freedom to experiment are hard-wired to be immediately attractive in young people. They want to be able to reveal all—and claim that they don't want anyone to invade their privacy—but in that dissonance, the latter interest is pushed aside. Embracing technology in the headlong pursuit of that freedom, the denizens of the social networks have discarded the concepts of privacy as a value and solitude as a personal good.

TECHNOLOGY MEETS ADOLESCENCE AND YOUNG ADULTHOOD

> "The highest Tax was upon Men who are the greatest Favourites of the other Sex, and the Assessments, according to the Number and Natures of the Favours they have received; for which they are allowed to be their own Vouchers."
>
> JONATHAN SWIFT, ENGLISH AUTHOR[2]

"Yeah, I am naked on the Internet," Kitty Ostapowicz told *New York* magazine in early 2007. Ostapowicz, then a 26-year-old bartender in New York's East Village, said that at 35, with her photos just a Google search away, she'd be proud. "It's a documentation of my youth, in a way. It's myself, what I used to be, what I used to do."[3]

Ostapowicz, one of the early adopters of one of the earliest social media sites on the Web, documented her life, clothed and unclothed, on LiveJournal, an April 1999 San Francisco startup. Nudity, however, is just one form of the revelations commonly found on the site. More commonly, posts consist of diary-like confessions of regret over doomed-from-the-start relationships, depression over the death of a parent or angst about the future of the planet. Other LiveJournal members can post words of encouragement, reveal their own tales of woe, or just commiserate with strangers. LiveJournal, while not as wildly popular as Facebook, has nonetheless carved out a niche for itself among Internet exhibitionists and voyeurs.[4]

Other websites that allow unfiltered broadcasts from users' webcams include Stickam, Chatroulette and LiveLeaks, a French company, which have attracted thousands of new users since YouTube began policing content and responding aggressively to complaints. Where YouTube's web slogan is "Broadcast Yourself," Stickam's is "Broadcast Yourself LIVE!" That openness has created public relations problems, as a *New York Times* blogger reported in late 2009 that three Stickam users had been arrested on sex crime charges in the previous nine months.[5]

Society's embrace of technology may provide the ideal pathway for the creation of a public persona in the digital world, allowing everyone to become a publisher. Political scientists have noted that while the easy availability of communication devices acts as a democratic leveler, democratization of culture does not promote the private, but has just the opposite effect, tempting users to push ever more of their private lives into the public domain.[6] One of the major drivers of the renegotiation of the public and private divide is the convergence of digital technologies. Digital personae alter the way people present themselves online and the way they relate to strangers both online and off.[7]

Many Facebook users post the intimate details of their lives without much apparent thought, according to a series of 2009 studies conducted in Australia by Sophos, an Internet security firm. Sophos researchers found that 41 to 46 percent of users "blindly accepted" friend requests from two fictitious users created by the firm, which then accessed dates of birth, e-mail addresses and school attendance information on 89 percent of respondents. More than half also provided their city or town of residence, leading Sophos to warn that such revelations made "an excellent starting point for scammers and social engineers."[8] Perhaps young people feel they have no choice but to seek acknowledgement from their cultural tribe. As Rashid Tobaccowala, CEO of the new media consultancy Denuo, has said, "If you aren't posting, you don't exist. People say, 'I post, therefore I am.'"[9] Such attitudes and constant activity have moved social media to the top spot in Internet search activity, where it comprises 20 percent of online traffic, according to online data researcher Bill Tancer.[10] The popularity of social media has helped push the long-time leader in Internet searches—porn—into second place at 10 percent.

Being, above all, seen, abetted by Internet technology, is now a shortcut to micro-celebrity status unrelated to ability. Being noticed is the goal, televised the pinnacle of achievement, opening the way for the least-talented to become public figures, according to Hal Niedzviecki.[11] When a teen comes out of nowhere and lands a bit part in a 30-second television commercial based on an Internet video clip, anything seems possible, so why not? Whether celebrity culture drives, or is driven by, narcissism is open to debate, however researchers find a significant generational shift in self-regard among young people.

Conducting a meta-analysis of 85 samples of college student answers to the 40-item Narcissism Personality Inventory between 1987 and 2006, San Diego State psychology professor Jean M. Twenge and four colleagues noticed that over that period, mean scores increased correlated to the year of date collection, so that by 2006 the mean score placed two-thirds of students above the 1985 mean, an increase of 30 percent.[12] Narcissism tracks closely with fame, and while young Americans increasingly want to become famous, large percentages also think it likely that they will become famous. In a Harris Poll conducted in 2000, 59 percent of people in their late twenties thought it at least somewhat likely that they would be at least briefly famous, even higher than the 44 percent of 18- to 24-year-olds who felt that way. Nearly a third of those between 30 and 49 also clung to that belief.[13]

Moreover, new research has shown that the rewards of narcissism are not only psychological, but also physical, and social media play a large role in providing a direct route to the brain's reward centers. Harvard researchers have recently concluded that social media are literally addictive because they tap into a hardwired human instinct to tell others about ourselves. The researchers found that up to 40 percent of human communication is devoted to describing our own subjective experiences and that self-disclosure lit up the same parts of the brain that are activated by having sex, eating favorite foods and, to a lesser extent, earning money. They found that the neurochemical reward was so strong that people turned down small sums of money offered for changing the subject, preferring to remain the subject of the conversation. The neurological activity was even greater when participants were able to share their thoughts with multiple people, they found, helping explain why about 80 percent of Facebook and Twitter posts consist of people writing about their own thoughts, opinions, and experiences.[14]

Other psychologists have previously explained self-disclosure activity with reference to the "disclosure decision model" which posits that disclosures are intended to achieve some goal, such as social approval, intimacy, self-fulfillment, blowing off steam or saving time in communication. In the process, there is a constant evaluation of risks and rewards such as reciprocity, extension of life off-line, and trust-building. Each possibility creates an opportunity and, possibly, a need to create a different social identity, each of which dictates how the user presents herself to the rest of the world.[15]

That identity usually starts with the user's screen shot, often a self-portrait selected from a variety of poses—from smiling to sultry—created with the camera built into virtually every cellphone. Equipped with a virtual digital darkroom, the technology coupled with self-absorption has spawned a new genre of photography which helps establish, position and curate their "brand," by allowing users to change photos at whim, expressing a teenage impulse that developmental psychologists call the "imaginary audience." The term, which is used to describe the motivation for a variety of adolescent behaviors, has been described as the "tendency among adolescents to over magnify the attention that others pay to their behavior and tend to imagine that they are constantly the object of attention of a rapt audience."[16]

From blogs, where even the most personal and trivial information is uploaded for public consumption to the new folk art of self-referential photography, it's a short step (via a link to YouTube) to the entirely new video form known as "haul" videos, in which people display, describe and generally show off things—often with the tags still on—they recently bought. In an earlier time, such ostentatious public airings would be derided as crass, tacky, and worse, but as a niche form made possible solely by the existence of YouTube, it is now viewed as an avant-garde public art form.

For the hyper-connected, services such as Twitter, Kyte, Radar, Helio, and Jaiku let young people stay connected at all times through a constant barrage of content sent out to an ever-shifting group that could number dozens to millions at any given time. Users don't know the size or composition of the audience, but then again, that's not the point. "As long as someone is connected, watching," Kyte co-founder Daniel Graf explained.[17]

But what if you can't manage even the basics of home videography and you have to fall back on words? And words fail you, too? The need to express yourself burns, but no ideas come to mind, even a 140-character "tweet" for your followers on Twitter. You need Plinky, an online tool to stimulate interesting musings that you can feed to any of your accounts that go out to the world or just beef up your blog for the edification of your readers. Plinky suggests that you discuss your best idea for surviving the zombie apocalypse. No matter how meaningless, Plinky's idea is to get you to share information about yourself, not just often, but constantly.[18]

Newer applications allow users to move from broadcasts to undifferentiated mass audiences from vaguely identified places to broadcasts to specific people from identified locations, shifting the privacy dynamics in subtle but important ways. The idea started in comparatively primitive fashion in 2004 with a service called Dodgeball, which allowed users to send an e-mail blast telling recipients where the sender was. Later, taking advantage of the next generation of mobile technology, "check-in" services such as Foursquare, Loopt, Yelp and Google's Latitude, use the built-in GPS capabilities of modern cell phones to notify friends of their locations, which serve as an invitation to join the user there.[19]

Despite these misgivings, several academics are convinced that data collection devices will be ubiquitous in the not-too-distant future and are working on ways of triangulating the data points collected by all the various personal electronic devices in existence or under development. "I fully believe that we will all be wearing this stuff all the time," according to Mark Bolas, a visiting professor in the University of Southern California film school. For those who demur, "The day before you die, your kids are going to look at you, when everyone else is doing this and say, 'You mean you didn't record when you were growing up? You're going to die and [your history] is going to go away?'" he explained. To cut through the quotidian clutter, Daniel Ellis, a professor of electrical engineering at Columbia University, is working on a method of homing in on particular clips from the whole, highlighting privacy issues further while also erasing the cultural value of forgetting and remembering to the development of the maturing individual.[20]

Technical developments designed to tease useful bits from a recorded life aside, some individuals have themselves developed an acute appreciation for the invasive possibilities presented by appearing on someone else's computer screen, however fleetingly. Recently, Steve Mann, a professor in the department of electrical and computer engineering at the University of Toronto, claimed he was assaulted by employees of a Paris McDonald's who tried to remove his "digital eyeglasses" pursuant to the chain's "no cameras" rule. Mann has worn some version of the device, which is of his own design and physically attached to his skull, since the 1980s. It records everything he sees, like Google's Project Glass, and it recorded the July 1 McDonald's encounter, which McDonald's described as "polite." Mann's images, which create a somewhat different picture of events, are posted, unsurprisingly, on his blog.[21]

Whereas "lifelogging" may be a generalized manifestation of a generational attitude of "letting it all hang out,"[22] the recent propensity to record and share extends to what not long ago was considered one of parenthood's most private moments. Often supplementing—if not replacing—the baby shower, the gender-reveal party may be the next logical step from announcing a pregnancy via social media. Today's parents, who themselves were the first generation whose own parents learned their child's sex in the obstetrician's office by ultrasound, have moved the moment even beyond the examining room.[23]

Now, many expectant parents share the exact moment at which they learn their baby's gender by opening the ultrasound test results (or cutting into a secretly baked pink or blue cake covered with plain white frosting) in the company of friends and relatives—then often posting the video on YouTube. The first gender-reveal video was posted on YouTube in 2008 and a "handful" were also uploaded in 2009 and 2010. However the genre took off in October 2011, with more than 1,800 appearing over the next six months. Lacking the drama of a video but not the popularity, gender-reveal discussion threads on BabyCenter.com, which draws 11 million visitors a month, have popped from 28 to 282 since early 2011.[24]

Not all revealing videos center on such festive events, nor are they as innocuous. For a variety of reasons, the temptation to send sexually explicit photos from smartphones to each other—and occasionally many others—is more than many teens can resist. Sometimes, the exchange of nude photos is a form of courtship; sometimes it is a form of revenge following a quarrel. In many of these cases, senders have run afoul of felony child pornography laws when prosecutors fear they may make their way onto the Internet and into the hands of sexual predators. Since 2009, at least 20 states have passed criminal statutes punishing sexting by minors as alarmed legislators try to rein in the technology and adolescent behavior in one fell swoop.[25]

Concerns about sexting and related behaviors may be highlighted by findings that significantly more adolescent girls than boys are active online. Among girls 15 to 17, 70 percent have a social media site profile, while only 57 percent of boys the same age do.[26] In a recent study by researchers at the University of Utah, only 5.1 percent of high school girls said that sending, receiving, or forwarding nude photos was acceptable, while 12.2 percent of boys had the same opinion. While girls condemn sexting by a wide margin, actual behavior tells a different story, with 17.3 percent of girls saying they have actually sent a nude photo, trailing boys by only 1 percent.[27]

Perhaps in recognition of the realities of teen communication styles, some states are considering modifying their laws to treat sexting teens differently than adult pornographers, making sexting a misdemeanor offense or even a form of truancy. "Generally, this should be an education issue," said Witold Walczak, legal director of the Pennsylvania chapter of the American Civil Liberties Union. "No one disputes that sexting can have very bad consequences, and no parent wants kids sending out naked images. But if you've got thousands of kids engaging in this, are you going to criminalize all of them?"[28]

WHICH WAY PRIVACY?

"I like my privacy as well as the next one...."
HUGO L. BLACK, ASSOCIATE JUSTICE, U.S. SUPREME COURT[29]

As a Harvard Law School professor and scion of a prominent New England family, Samuel Warren was used to traveling in the rarified social circle of the Boston Brahmins. What he was not used to was reading about it in the newspaper. Shortly after his 1883 marriage to Mabel Bayard, Warren's law practice prospered and the young couple began to entertain lavishly at their Beacon Hill home.[30] This attracted the attention of the *Saturday Evening Gazette*, which reported the events in detail.[31]

Warren and his friend and former law partner, Louis Brandeis, who would join the U.S. Supreme Court in 1916, confronted the press' intrusion into private life in their seminal law review article, "The Right to Privacy," in which they defined the right of privacy as "the right to be let alone." With the depredations of the *Gazette* clearly in mind, the two complained that "Instantaneous photographs and newspaper enterprise have invaded the sacred precincts of private and domestic life…. The press is overstepping in every direction the obvious bounds of propriety and of decency. Gossip is no longer the resource of the idle and the vicious, but has become a trade, which is pursued with industry as well as effrontery…. To occupy the indolent, column upon column is filled with idle gossip, which can only be procured by intrusion upon the domestic circle."[32]

The first two cases citing the Warren and Brandeis article as support for privacy rights were brought by plaintiffs whose photos had been used in advertising without permission. New York state judges in both cases ruled that the authorities Warren and Brandeis relied on were too remote to be persuasive and denied the privacy claims.[33] Though the courts were slow to adopt the concept of privacy, the public and their state legislators were not and privacy statutes were soon widely adopted. On the Supreme Court, Brandeis continued to have influence over the development of privacy as a right, notably in his dissent from the Court's holding in *Olmstead v. United States* refusing to treat wiretapping as a Fourth Amendment violation.[34]

Commenting on the purposes of the Fourth and Fifth Amendments, Brandeis wrote: "…The makers of our Constitution undertook to secure conditions favorable to the pursuit of happiness. They recognized the significance of man's spiritual nature, of his feelings and his intellect... They sought to protect Americans in their beliefs, their thoughts, their emotions and their sensations. They conferred, as against the Government, the right to be let alone—the most comprehensive of rights and the right most valued by civilized men."[35]

Of course, Warren and Brandeis were not the first to recognize the value of the private sphere. Aristotle made the distinction between the public sphere of political participation (the polis) and the private domestic sphere of the family (the oikos). In the mid-19th Century, the right to be free from the interference of others in one's personal activities was famously outlined by John Stuart Mill in *On Liberty*, where he addresses the issue twice. In the early part of the essay Mill writes, "The only part of the conduct of any one for which he is amenable to society, is that which concerns others. In the part which merely concerns himself, his independence is, of right, absolute."[36] In his chapter on the authority of the state, he continues this theme: "As soon as any part of a person's conduct affects prejudicially the interests of others, society has jurisdiction over it.... But there is no room for entertaining any such question when a person's conduct affects the interests of no persons besides himself..."[37]

Such tender sensibilities seem unlikely to be asserted in the United States in the early 21st Century. In reviewing the Warren and Brandeis article and the Warrens' complaints which were its genesis, Stewart Baker asserts that privacy as a concept is passé. "Is there anyone alive who thinks it should be illegal for the media to reveal the guest list at a prominent socialite's dinner party? Today it's more likely that the hostess... will blog about it in advance, and that the guests will send Twitter updates while it's under way."[38]

While Americans claim to value privacy in the abstract, their online behavior would suggest otherwise. And when privacy concerns do emerge, they tend to be relatively short-lived. In a 2008 study by AOL of Internet users, 84 percent of respondents said they would not disclose details about their income online—but 89 percent of that group then did just that.[39] Two years later, researchers found that in the year following heated discussions about changes to Facebook's privacy policies, modifications to privacy settings increase among 18- and 19-year-olds, but then tail off sharply.[40] Similar reactions are also noted when Facebook introduces new services, such as Beacon, which notified users' "friends" (without the users' consent) about online purchases.[41]

While Europeans have embraced the new communication technologies with equal enthusiasm, they have also been slower to loosen their stricter laws and social habits governing privacy than Americans. Americans were understandably outraged when a video of four middle-school boys verbally abusing a 68-year-old grandmother who was a volunteer monitor on their school bus was posted on YouTube.[42] The boys were suspended from school for a year, but no further legal action was taken. By contrast, in Italy, three Google executives received suspended six-month jail terms for allowing a video of an autistic boy being bullied by fellow students to be posted on their site and then remain up for two months.[43]

Europeans have historically regarded privacy as a fundamental human right, a view reinforced by living memories of totalitarian regimes that depended on a web of surveillance and informers to cultivate a climate of fear that allowed them to hold onto power. As a result of that experience, Europeans have enshrined privacy in the European Convention on Human Rights, where it occupies a more prominent position than press freedom.[44] American law professors Jack Goldsmith and Tim Wu, who study global press law, have written, "For many purposes, the European Union is today the effective sovereign of global privacy law."[45]

Since 2002, Monaco's Princess Caroline has waged a privacy battle against German photo magazines that have published rather tame photos of her going about her daily activities. Although the European Court of Human Rights recently ruled in the magazine's favor, it placed strict limits on the press' right to report on the private lives of public figures.⁴⁶ In 2004, judges of the same court had found in the princess' favor. In a widely quoted concurring opinion from that case, Judge Bostjan Zupancic of Slovenia wrote, "The German privacy law doctrine of Personlichkeitsrecht (individual rights) testifies to a broader concentric circle of protected privacy. I believe that the courts to some extent and under American influence made a fetish of the freedom of the press. It is time that the pendulum swung back to a different kind of balance between what is private and secluded and what is public and unshielded."⁴⁷

Whichever way the legal landscapes of privacy shift on each side of the Atlantic, Europeans will still look aghast at American attitudes. Their devotion to the private goes only so far, however, and like gawkers at an accident, Europeans relish the irony of being hoist with one's own petard. According to France's third-largest daily, the center-left *Le Monde*, only a week after Facebook adjusted its privacy options so that some material would be treated as public by default, founder Mark Zuckerberg discovered that nearly 300 photos on his personal site that had been private were available to all. After a selection of the photos appeared on other websites, they quickly disappeared from Zuckerberg's personal site, the newspaper noted with some delight.⁴⁸

IT'S ALL BUSINESS

> "I think that there is nothing, not even crime, more opposed to poetry, to philosophy, nay, to life itself than this incessant business."
>
> HENRY DAVID THOREAU, AMERICAN PHILOSOPHER⁴⁹

When Mark Zuckerberg announced in early April 2012 that Facebook would finance a $1 billion deal for the photo-sharing service Instagram with only $300 million in cash and the rest in stock, the social media giant gave investors a peek at the company founder's view of the value of its own stock. Instagram, a relative newcomer to the social media ranks, had no revenue and 12 employees, but had been valued at roughly $500 million by a group of venture capitalists just days before. At that level, Zuckerberg was pegging Facebook shares at about $30, which would value the company at just over $80 billion.⁵⁰

In the run-up to Facebook's IPO, most analysts expected the company to make its public debut at between $28 and $35 a share, valuing the company at $78 to $96 billion. However, two days before the IPO, Facebook announced that it was increasing its initial offering by 100 million shares, or 25 percent. Simultaneously, some insiders announced that they would sell large chunks of their holdings as soon as the three-to-seven-month sales restriction (the "lock-up" period) expired. News of the increased number of shares, coupled with a decision to raise the offering price as high as $38, was read as an indication of heavy demand and boosted the valuation as high as $104 billion.⁵¹

On Friday, May 18, its first day as a NASDAQ public company, Facebook did indeed come out at $38, a price settled on after Zuckerberg's two-week nationwide "road show" meetings with institutional investors.[52] The stock's performance that day, however, was a stunning disappointment, closing up a miniscule 23 cents, when 10–15 percent is typical. The news was worse on Monday, when investors drove the price down almost 11 percent to barely over $34 and paring its market value to $93 billion. The company, its investment bankers at Morgan Stanley, and the NASDAQ were immediately criticized for difficulty in executing trades and the decision to increase the offering and its price, meaning that its price-to-earnings ratio of 85 seriously overvalued the company, based on projected 12-month earnings. By comparison, Google's current p/e ratio is about 13.[53]

Compounding its revenue problems, which are expected to reach only $1.1 billion in 2012, or just over $1 per registered user, are concerns about Facebook's mobile strategy. Users are turning to mobile access at an explosive rate, but games can't be played on Facebook's mobile site, and the social media company has only just begun experimenting with mobile ads.[54] Nearly simultaneous with Facebook's IPO another serious revenue problem surfaced when General Motors, the nation's third-largest advertiser, announced that it would pull its $10 million in advertising from the site, saying that the ads didn't sell cars.[55]

The dreams of a piece of the billions of dollars circulating through the universe of social media are certainly driving the proliferation of most, if not all, of the 3,000 new apps launched in just the first four months of 2012 that integrate with Facebook.[56]

And, just as certainly, that money is a vital part of the effort to bring the ideas behind them to fruition. But just as necessary an ingredient is the one most likely to be overlooked—the hardware platform that supports the software necessary to bring the apps' capabilities to users. As an answer to a 21st-Century version of the chicken-and-egg question, it appears reasonable to conclude that the development and constant improvement of handheld devices and the wireless network that ties everyone together stimulates the growth of content for them to carry.

It is just these types of advances in technology that so bedeviled Warren and Brandeis in the late 19th Century. "Solitude and privacy have become more essential to the individual; but modern enterprise and invention have, through invasions upon his privacy, subjected him to mental pain and distress," they wrote. "In this, as in other branches of commerce, the supply creates the demand." That demand eventually runs up against boundaries of privacy and our private matters are suddenly aired in the public sphere.[57]

As what were once primarily cellular telephones began to appear with Global Positioning System (GPS) capabilities, services began to appear linking users with businesses based on location and digital devices became primarily what their detractors pejoratively term "tracking devices." That term highlights what privacy advocates fear: who will have access to where we are, both right now and in some distance in the past? Are the data secure from hackers who would do us harm if they could find us easily? Do these digital footprints tempt shortcuts by law enforcement that amount to warrantless searches in violation of the Fourth Amendment? How long is location data stored before being purged?

Early on, privacy concerns centered around the problem of confining a user's location to a pre-selected circle of friends. One of the earliest location services, Boost, a unit of Sprint, urged users to include only close, trusted friends.[58] Today, smartphones contain sensors that are always connected to the Internet, automatically determining exactly where you are, which often allows anyone with access to also know exactly what you are doing. Connecting with a social network that automatically updates your friends on what music you're listening to has become known as "frictionless sharing." But when it collects data on web surfing habits and the like that is not uploaded to a social media site, it can become "frictionless surveillance." When used in conjunction with a location service, it is even possible to correlate your location data with that of your friends to predict where you will be in a few minutes or hours.[59]

CONCLUSION

To modernists, society was an oppressive regime, with the masses hostile and aggressive toward the self. This is the post-war attitude succinctly captured by French modernist playwright and philosopher Jean-Paul Sartre's memorable line: "Hell is other people."[60] If literary critic Lionel Trilling is correct that the defining characteristic of modernism is authenticity, then that of post-modernism is connectivity.[61] No longer oppressed by the masses, post-modernists fear being separated from them and embrace the artificial closeness provided by the very small screens of mobile devices.

As we've seen, the rush toward 24/7 connectivity and the accompanying revelations of every thought and interest have occurred with little, if any, forethought. Left behind and buried by the technologically enabled mobility, capability and excitement generated by the tsunami of interesting material to see and respond to, privacy has been lost before most users even thought about having had it. The practices, technologies and markets for personally identifiable information are developing faster than social norms that control how people protect information about themselves. Eventually, people will have to start creating traditions, values, and rules that will protect privacy and emphasize personal dignity and social fairness in a constantly connected electronic social universe—that is, what it means to be human in an information economy and an electronic social milieu. Perhaps it is the illusion of control, i.e. social media users decide what to post, that encourages sharing personal information with those people who users have "accepted" as friends. An altogether higher level of thought is required to grasp the interconnectedness of websites and understand the technical capabilities of sophisticated data collection programs to make those connections and crawl through that thicket of information to assemble detailed personal profiles enhanced by similarities which those users communicate with.

Concerns about the amount of personal information provided by users are often allayed by the possibility that there will be a reward for having done so. Surveys show that when people are in a "transactional" frame of mind regarding their personal information, they will share it more readily if they think they can get something of value in return.[62] Similarly, on-line retailers such as Amazon who emphasize a value-added benefit to data collection generally have few complaints from their loyal book-buying customers. It is not necessary to read Amazon's privacy policy to determine that the web site's visitors' book searches are tracked and stored. Customers don't object because they believe that the well-known feature that presents books-you-might-like is an up-front use of their information to help them make more informed buying decisions. If users come to believe that their personal information is being used "for" them rather than "against" them they are less likely to consider the implications of that information being tracked and sold.[63]

The difficulty—and challenge—for social media users, especially the youngest, will be to find a balance between access to what we want and control of what we have to give up to get it. Already, social media users can perceive the difficulties of switching away from the sites and services where their friends are and the near-impossibility of dealing with the hundreds of commercial entities holding personal data. No one born into the digital generation has yet lived into adulthood—let alone experienced the yet-to-be-revealed ramifications of having lived an aggregated digital life of 85 years or more.[64]

We've experienced the benefits of railroads and oil exploration, but many of the 19th- and early 20th-Century industrialists who brought those projects to fruition are still regarded as rapacious businessmen and are often referred to as "robber barons" more than 100 years later. Then, their least savory business practices led to tight federal regulation by vigilant Congressional watchdogs. This is not unlike the environment 21st-Century social media entrepreneurs are operating in. Their promises of the benefits of "liking" everything and connecting with everyone can't act as a smokescreen for base commercial motives that conflict with the broader interests of individuals and society as a whole. Surrendering privacy interests in exchange for entry into imagined "communities" is a bargain that requires close examination and robust discussion by a technologically savvy and media-literate citizenry. Such a national conversation may be required if we want to oppose the "surreal world of Web 2.0 where the rhetoric of democracy, freedom, and access is often a fig leaf for anti-democratic and coercive rhetoric; where commercial ambitions dress up in the sheep's clothing of humanistic values; and where, ironically, technology has turned back the clock from disinterested enjoyment of high and popular art to a primitive culture of crude, grasping self-interest."[65]

When people lived in tribes and villages, which were the basic living arrangements for most of human history, everyone knew everyone else (and were related to many of them) and knew generally what they did on a daily basis. Even within that small group, however, no one could learn everything about everybody and certainly could not recall every detail of dates, exact locations and conversations, let alone what they were thinking about. Even what we do learn is thinned out by time and imperfect memory. Technology, though, allows the collection and storage of every bit of data it searches out. And the databases never forget. During the social and technical revolutions that led us from close communal social relationships to the complex and widely dispersed populations of today, we've come to expect more from technology and less from each other. Traditional social ties have been replaced by illusions of companionship without the messy complexities of actual relationships. At the same time, the "data ties" that fuel social media and tempt advertisers to learn as much about us as possible make the right to be let alone in the digital age an urgent national issue.

NOTES

1. Polly Sprenger, "Sun on Privacy: 'Get Over It,'" *Wired*. January 26, 1999.
2. Jonathan Swift, *Gulliver's Travels*, Part III, Chapter VI, (1726, New York, NY: Oxford University Press, 1977).
3. Emily Nussbaum, "Say Everything," *New York Magazine*, February 19, 2007, 22.
4. Clint Boulton, "Six Apart Sells LiveJournal to Russia's SUP." *eWeek*, December 3, 2007. http://www.eweek.com/c/a/Messaging-and-Collaboration/Six-Apart-Sells-LiveJournal-to-Russias-SUP
5. Brad Stone, "Three Sex Crime Arrests Among Stickam.com Users So Far This Year," *New York Times*, October 15, 2009.
6. Andrew M. Koch and Amanda G. Zeddy, "The Rise and Fall of the 'Private' as Part of Western Political Socialization," *Politics, Culture and Socialization* 1 (July 2010): 1–29.
7. Amparo Lasen and Edgar Gomez-Cruz, "Digital Photography and Picture Sharing: Redefining the Public/Private Divide," *Knowledge, Technology & Policy* 22 (September 2009): 205–15.
8. Don Reisinger, "Study: Facebook Users Willingly Give out Data," *Webware*, December 7, 2009. http://news.cnet.com/8301-17939_109-10410257-2.html
9. Bob Garfield, "YouTube v. Boob Tube." *Wired*, December 2006, 12.
10. Bill Tancer, *Click: What Millions of People Are Doing Online and Why it Matters* (New York, NY: Hyperion, 2008): 19.
11. Hal Niedzviecki, *Hello. I'm Special: How Individuality Became the New Conformity* (San Francisco: City Lights Books, 2006): 17–19.
12. Jean M. Twenge and Joshua D. Foster, "Egos Inflating Over Time: A Cross-Temporal Meta-Analysis of the Narcissistic Personality Inventory," *Journal of Personality* 76 (December 2008): 875–901.

13. Harris Poll #43, Aug. 9, 2000.
14. Diana Tamir and Jason P. Mitchell, "Disclosing Information About the Self is Intrinsically Rewarding," *Proceedings of the National Academy of Sciences* 109 (May 2012): 8038–43.
15. John Palfrey and Urs Gasser, *Born Digital: Understanding the First Generation of Digital Natives* (New York, NY: Basic Books, 2008): 24–26.
16. Jeffrey J. Arnett, "Contraceptive Use, Sensation Seeking and Adolescent Egocentrism," *Journal of Youth and Adolescence* 19 (April 1990): 171–180.
17. Brad Stone and Matt Richtell, "Social Networking Leaves Confines of the Computer," *New York Times*, April 30, 2007, B4.
18. Rob Walker, "Say What?" *New York Times Magazine*, May 31, 2009, 30.
19. Jenna Wortham, "Telling Friends Where You Are (or Not)," *New York Times*, March 14, 2010, B6.
20. Scott Carlson, "On the Record, All the Time," *The Chronicle of Higher Education*, February 9, 2007, A31.
21. Angela Chen, "Professor Says He Was Assaulted Over Wearable Computer Glasses," *The Chronicle of Higher Education*, July 19, 2012.
22. The Hombres, "Let it All Hang Out," *Let it Out*, written by Gary McEwen and B.B. Cunningham, Verve Forecast Records (1967).
23. Ian Donald, John McVicar, and Tom Brown, "Investigation of Abdominal Masses by Pulsed Ultrasound," *The Lancet*, June 7, 1958, 1188–1195.
24. Alex Williams and Kate Murphy, "A Boy or Girl? Cut the Cake," *New York Times*, April 8, 2012, ST2.
25. National Conference of State Legislatures research report, "2012 Sexting Legislation," December 14, 2012. http://www.ncsl.org/issues-research/telecom/sexting-legislation-2012.aspx
26. Pew Internet and American Life Project, "Teens and Social Media," December 19, 2007.
27. Donald S. Strassberg, Ryan K. McKinnon, Michael A. Sustaita, and Jordan Rullo, "Sexting by High School Students: An Exploratory and Descriptive Study," *Archives of Sexual Behavior* 42 (January 2013): 15–21.
28. Tamar Lewin, "States Revise Porn Laws for Sexting Teens," *New York Times*, March 21, 2010, A1.
29. Hugo L. Black, dissenting in *Griswold v. Connecticut*, 381 U.S. 479, 510; 85 S.Ct. 1678, 1695–6 (1965): 510.
30. Mabel Bayard was a daughter of Thomas Bayard, who served as secretary of state (1885–1889) under President Grover Cleveland.
31. The *Gazette* (1851–1906) was a successful Boston weekly typical of the "yellow journalism" era of the late 19th century. Representative issues of the *Gazette* from the 1880s are held by the Boston Public Library and the Massachusetts Historical Society, Boston, Mass.
32. Samuel D. Warren and Louis D. Brandeis, "The Right of Privacy," *Harvard Law Review* 4 (December 1890): 193.
33. *Roberson v. Rochester Folding Box Co.*, 171 N.Y. 538 (1902); *Moser v. Press Pub. Co.*, 109 N.Y.S. 963 (1908).
34. *Olmstead v. United States*, 277 U.S. 438 (1928).
35. *Olmstead v. United States*, 478.
36. John Stuart Mill, *On Liberty*, ed. Alburey Castell (1859, New York, NY: Appleton-Century-Crofts, 1947): 10.
37. Mill, *On Liberty*, 75–76.

38. Stewart A. Baker, *Skating on Stilts: Why We Aren't Stopping Tomorrow's Terrorism* (Stanford, CA: Hoover Institution Press, 2010): 311.
39. Gordon L. Crovitz, "Privacy? We Got Over It," *Wall Street Journal*, August 25, 2008, A18.
40. Dana Boyd and Eszter Hargittai, "Facebook Privacy Settings: Who Cares?" *First Monday* 15 (August 2010).
41. Adam Cohen, "One Friend Facebook Hasn't Made Yet: Privacy Rights," *New York Times*, February 18, 2008, B16.
42. Charles M. Blow, "Bullies on the Bus," *New York Times*, June 23, 2012, A19.
43. Rachel Donadio, "Larger Threat is Seen in Google Case," *New York Times*, February 25, 2010, A1.
44. European Convention on Human Rights, Council of Europe Treaty Series, No. 5. Strasbourg, France, 2012: 10–12.
45. Jack Goldsmith and Tim Wu, *Who Controls the Internet?* (New York, NY: Oxford University Press, 2006): 174.
46. Wil Longbottom, "Media Interest in Lives of Celebrities is Legitimate, Human Rights Judges Rule," *Daily Mail*, February 8, 2012, 1.
47. *von Hannover v. Germany*, No. 59320/00, 2004-VI, ECHR, 294.
48. "Le Fondateur de Facebook Piege par les Nouveaux Parameters 'vie Privee' du Site (Facebook Founder Trapped by Site's New Privacy Settings)," *Le Monde*, December 14, 2009, 22.
49. Henry D. Thoreau, "Life Without Principle," *Atlantic Monthly*, 12 (September 1863): 484–95, 485.
50. Evelyn M. Rusli, "Talks With Instagram Suggest a $104 Billion Valuation for Facebook," and "The Chatter," *New York Times*, April 15, 2012, BU2.
51. Shayndi Raice, Anupreeta Das, and Lynn Cowan, "Facebook Insiders Boost Plans to Cash Out in IPO," *Wall Street Journal*, May 17, 2012, A1-2.
52. Shayndi Raice, Anupreeta Das, and John Letzing, "Facebook Prices IPO at Record Value," *Wall Street Journal*, May 18, 2012, A1.
53. Jacob Bunge, Aaron Lucchetti, and Gina Chon, "Investors Pummel Facebook," *Wall Street Journal*, May 22, 2012, A1-2.
54. Shayndi Raice, "Facebook's Mobile Miscalculation," *Wall Street Journal*, May 22, 2012, B1.
55. Sharon Terlep, Suzanne Vranica, and Shayndi Raice, "GM Says Facebook Ads Don't Pay Off," *Wall Street Journal*, May 16, 2012, A1.
56. *CBS Evening News*, May 19, 2012.
57. Warren and Brandeis, "The Right of Privacy," 196.
58. Randall Stross, "Cellphone as Tracker: X Marks Your Doubts," *New York Times*, November 19, 2006, BU3.
59. Peter Maass and Megha Rajagopalan, "That's No Phone. That's My Tracker," *New York Times*, July 13, 2012, SR5.
60. Jean-Paul Sartre, *No Exit and Three Other Plays* (New York, NY: Vintage, 1989).
61. Lionel Trilling, *Sincerity and Authenticity* (Cambridge, MA: Harvard University Press, 1972).
62. Joshua Brustein, "Tag-Along Marketing," *New York Times*, November 7, 2010, WK2.
63. Quentin Fottrell, "10 Things Amazon Won't Tell You: 8. 'We Know More About You Than You Think,'" *Smart Money*, August 2012, 102.
64. Palfrey and Gasser, *Born Digital*, 62.
65. Lee Siegel, *Against the Machine: Being Human in the Age of the Electronic Mob* (New York, NY: Spiegel & Grau, 2008).

Editors and Contributors

Charles N. Davis is Dean of the Henry W. Grady School of Journalism and Mass Communication at the University of Georgia. He spent the past 14 years as a professor at the University of Missouri School of Journalism. Davis worked for ten years as a journalist after his graduation from North Georgia College, working for newspapers, magazines and a news service in Georgia and Florida before leaving full-time journalism to complete a master's degree from the University of Georgia's Henry W. Grady School of Journalism and Mass Communication and to earn a doctorate in mass communication from the University of Florida. His teaching awards include the Scripps Howard Foundation National Journalism Teacher of the Year Award in 2008, the Provost's Award for Junior Faculty Teaching in 2001, and the University of Missouri Alumni Association's Faculty/Alumni Award in 2008.

David Cuillier is an associate professor and director of the University of Arizona School of Journalism. He was a newspaper reporter and editor in the Pacific Northwest before earning his doctorate at Washington State University in 2006. At the University of Arizona he teaches and researches access to government records and is co-author with Charles N. Davis of another book, *The Art of Access: Strategies for Acquiring Public Records*. His researchfocuses on public attitudes toward access, the implementation and effectiveness of freedom of information laws, and the psychology of access—the human interaction between requester and agency. He has testified before Congress regarding the Freedom of Information Act and he provides training and education to journalists and the public in accessing public records. For four years he served as Freedom of Information Committee chairman for the Society of Professional Journalists (SPJ), the most broad-based organization of journalists in the United States, and he served as national president of SPJ in 2013–14.

Sigman L. Splichal is an associate professor in the Department of Journalism and Media Management in the School of Communication at the University of Miami. A journalist for more than 20 years before pursuing an academic career, Dr. Splichal's research and teaching interests focus on communication law, ethics, and professional practice. In particular, he's interested in how technology is altering the traditional privacy laws and perspectives. He is the former director of the Journalism program at the University of Miami and currently is the program's graduate coordinator.

Martin E. Halstuk teaches communications law and also news reporting and writing at The Pennsylvania State University. Dr. Halstuk is Senior Fellow at the Pennsylvania Center for the First Amendment, which assists journalists, scholars, and others in First Amendment issues and in obtaining access to government information. Before turning to teaching and legal research, he worked as a newspaper reporter and editor for 21 years. He has been a copy editor at the *Los Angeles Times*, night city editor at the *San Francisco Examiner* and a courthouse reporter for the *San Francisco Chronicle*. Over the years, he has received top awards for his work as a journalist as well as his legal scholarship. He has also taught at UCLA, the University of San Francisco, St. Mary's College of California and the University of Nevada. His publications on press and public access to government-held information have been cited in media *amici* briefs presented in cases argued before the U.S. Supreme Court and several federal courts. His scholarly articles have appeared in more than a dozen law publications including *Stanford Law & Policy Review, William & Mary Bill of Rights Journal, University of Florida Journal of Law and Public Policy*, and *Communication Law and Policy*.

Joey Senat is an associate professor at Oklahoma State University's School of Media and Strategic Communications, where he teaches mass communication law and multimedia journalism courses. Dr. Senat writes an open government blog and is quoted frequently by national and statewide news outlets on freedom of information and media law issues. He has spoken on FOI, First Amendment and journalism education issues at dozens of professional and academic conferences. Dr. Senat serves on the FOI Committee of the national Society of Professional Journalists. For his work to advance government transparency, Dr. Senat received the 2007 Marian Opala First Amendment Award and the 2005 Oklahoma Society of Professional Journalists Award for Distinguished Service to the First Amendment. He has been named Outstanding Professor in OSU's College of Arts and Sciences and received the Mortar Board Honor Society's Golden Torch Award for the college. Prior to becoming a college professor, Dr. Senat was a reporter for *The Commercial Appeal* in Memphis, Tennessee, and the *Tulsa (Oklahoma) World*. He earned a bachelor's degree from Louisiana State University, a master's from Memphis State University and a doctorate from the University of North Carolina at Chapel Hill.

Cheryl Ann Bishop taught communication law at Quinnipiac University in Connecticut for five years and is the author of *Access to Information as a Human Right*. Conceptualizing access to government information as a human right is a new development in the global trend promoting institutional transparency. In *Access to Information as a Human Right*, Dr. Bishop provides a comprehensive examination of international human rights law and explains four conceptualizations of access to information as a human right. She concludes that a human right to access information is evolving in disparate ways. The current evolution of access rights creates a patchwork system of guarantees; nonetheless, the freedom-of-expression conceptualization holds the most promise for proving a broad right of access. Bishop received her Ph.D. in mass communication with an emphasis on international communication law at the University of North Carolina and a M.A. in journalism at the University of Missouri. She resides in Tucson, Arizona, and does communications and public affairs for an international nonprofit organization.

Richard J. Peltz-Steele, is a professor teaching in tort, media, and access law at the University of Massachusetts Law School. He received his law degree from Duke University and bachelor's in journalism and Spanish from Washington & Lee University. Peltz-Steele practiced commercial law in Baltimore, Maryland, before teaching, and has won awards in teaching, research, and public service, the latter for work in state freedom of information law. He is author or co-author of research articles in law and mass communication, U.S. and comparative, qualitative, and quantitative law, as well as book chapters, a treatise, and two casebooks and teaching manuals, including *The Law of Access to Government* from Carolina Academic Press. Peltz-Steele is active in the American Bar Association and has taught or presented papers throughout the United States and in Africa, Asia, and Europe. Last year he published *The New American Privacy*, a research article on proposed EU privacy regulation and U.S. privacy policy, in the *Georgetown Journal of International Law*. His current projects include liability in U.S. tort law for the publishers of leaked corporate secrets and the relationship between sport communication and international development.

Derigan Silver, is an associate professor and director of undergraduate studies in the Department of Media, Film and Journalism Studies and an adjunct faculty member in the Sturm College of Law at the University of Denver. He teaches courses on media law and policy, Internet law and policy, and the First Amendment. He is the author of numerous book chapters and articles on national security information law, defamation, access to court documents and proceedings, Internet law and policy, and First Amendment theory. His book, *National Security in the Courts: The Need for Secrecy vs. the Requirement of Transparency* was published in 2010. He received his bachelor's degree from the University of California at Santa Barbara, his master's degree from the Walter Cronkite School of Journalism and Mass Communications at Arizona State University, and his doctorate from the School of Journalism and Mass Communication at the University of North Carolina at Chapel Hill.

Kyu Ho Youm, Professor and the Jonathan Marshall First Amendment Chair at the University of Oregon School of Journalism and Communication, has published about communication law in a number of academic and trade journals in the U.S. and abroad. His law journal articles have been cited by American and foreign courts, including the U.K. House of Lords, the Australian High Court, and the Canadian Supreme Court. American and international lawyers have used his media law research in representing their clients in press freedom litigation. Youm has prepared a freedom of information report of 23,000 words for Open Society Justice Initiative's The Right to Information: Good Law and Practice. He has been involved as a libel law author in writing *Communication and the Law*, a U.S. media law textbook. Also, he has contributed to *International Libel & Privacy Handbook* and *Media, Advertising, & Entertainment Law Throughout the World*. A native of South Korea, he has authored a book on Korean press law. Youm is the communication law and media policy editor of the 12-volume *International Encyclopedia of Communication*. He studied journalism at Southern Illinois University-Carbondale for his master's and doctorate and law at Yale and Oxford for his graduate degrees.

EDITORS AND CONTRIBUTORS | 165

Jonathan Peters is a media lawyer and an assistant professor at the University of Dayton, where he teaches journalism and law. He blogs about free expression for the *Harvard Law & Policy Review*, and he has written on legal issues for *The Atlantic*, *Slate*, *The Nation*, *Wired*, PBS, and the *Columbia Journalism Review*. Peters is a volunteer attorney for the Student Press Law Center and the Reporters Committee for Freedom of the Press, both in Washington, D.C., and for the Online Media Legal Network at Harvard University. Peters is the First Amendment Chair of the Civil Rights Litigation Committee of the American Bar Association, a member of the Media Law Committee of the Ohio State Bar Association, and a member of the Board of Directors of the ACLU of Ohio. He has a journalism degree from Ohio University, a law degree from Ohio State University, and a doctorate in journalism from the University of Missouri.

Daxton R. "Chip" Stewart, is an associate professor in the Bob Schieffer College of Communication at Texas Christian University, where he has taught classes in media law and ethics since 2008 and has served as Associate Dean since 2013. Dr. Stewart began his journalism career as a sportswriter in Dallas before attending law school at the University of Texas. After practicing law in Texas, he became a city editor on the cops and courts beat and a columnist at the *Columbia Missourian* while earning his master's and doctorate in journalism and master of laws in dispute resolution at the University of Missouri. While in Missouri, he focused his research on freedom of information law, particularly on sanctions for public records and open meetings laws violations and on alternatives to litigation when disputes arise under open government laws. Dr. Stewart has published articles in journals such as *Journalism and Mass Communication Quarterly*, *Communication Law and Policy*, *American Journalism*, the *Journal of Media Law and Ethics*, the *Journal on Telecommunications and High Technology Law*, and the *Journal of Dispute Resolution*, and the *Appalachian Journal of Law*. He was editor of *Dispute Resolution Magazine*, the quarterly publication of the American Bar Association's Section on Dispute Resolution, from 2007 to 2012. He edited the book *Social Media and the Law: A Guidebook for Communication Students and Professionals* (Routledge, 2013), and he is the founding editor of *Community Journalism*, an online, open access, peer-reviewed journal that launched in 2012.

Paul Gates has been a professor on the faculty of the Department of Communication at Appalachian State University in Boone, North Carolina since 1995. He was awarded the doctorate in media law and policy from the University of Florida in 1996 and earned an M.C.C. from the University of South Carolina in 1980. He has been a member of the Louisiana State Bar Association since 1986. Gates is a former newspaper and wire service editor and reporter and won prizes for investigative reporting while a law student in San Diego in 1985. He has published numerous journal articles and book chapters on a variety of legal topics. His latest book (co-authored) is *Medieval America: Cultural Influences of Christianity in the Law and Public Policy*, published by Lexington Books in 2012. He also lectures frequently at professional conferences on issues of faculty governance and academic freedom.

Index

A

Abramson, Federal Bureau of Investigation v., 20
access
    to court proceedings, 85–87
    to court records, 67–70 *See also* electronic court record access
administrative adjudication, for dispute resolution, 138–39
Administrative Procedure Act of 1946, 67, 74
adolescence, social media and, 146–51
Alaska courts
    electronic media coverage in, 93
    on email as public records, 105
Alternative Dispute Resolution Act, 136
Amar, Akhil, 98
ambient awareness, 146
Angwin, Julia, 123
appellate courts, electronic media coverage in, 94
Aristotle, 152

Arizona
    court rulings on email privacy, 107
    ombuds office in, 138
Arkansas Democrat-Gazette, Pulaski County v., 107
Arkansas Supreme Court
    on bulk and compiled access, 72
    on FOI and email privacy, 107
Article 29 Working Party, 58–59
Asia Global Crossing Ltd., 102
Associated Press v. Canterbury, 107–8

B

Baker, Stewart, 153
Balkin, Jack, 122
Banisar, David, 51
Bayard, Mabel, 151
Bennett, Mark, 91
Bent, Bruce, II, 103
Bird, Rose Elizabeth, 38
Bishop, Cheryl Ann, 163

Black, Hugo L., 151
Blackmun, Harry, 22–23
Board of County Commissioners of County of Arapahoe, Denver Publishing Co. v., 107
Bolas, Mark, 150
Boost location service, 156
Brandeis, Louis, 10, 152, 155
Brennan, William J., Jr.
  on computer privacy issues, 6, 8
  on government transparency, 19–20
  on open courts, 86
broadcasting
  from federal courts, 87–91
  from state courts, 91–94
bulk and compiled electronic court record access, 71–72
burden of proof, for FOIA users, 24–26
Burger, Warren, 89
Bush, George W., 26, 27–30

C

California Court of Appeals
  computerized information issues and, 41–42
  FOI and email privacy and, 108–9
  workplace email privacy and, 103
California Supreme Court
  on bulk and compiled record access, 72
  on electronic media in courts, 92–93
cameras
  in federal courts, 87–91
  in state courts, 91–94
Canons of Professional and Judicial Ethics, 88
Canterbury, Associated Press v., 107–8
Caroline, Princess of Monaco, 153–54
Carter, Robert L., 36
categorical exemptions, to court record access, 72–74
CCJ/COSCA (Conference of Chief Justice/Conference of State Court Administrators) guidelines, 69–70, 72
central purpose standard, for FOIA, 22–24
Chandler v. Florida, 88–89
City of Clearwater, Florida v., 107

City of Monroe, Mechling v., 109
City of Ontario v. Quon, 99–100
civil trials, right of access and, 87
Clark, Tom, 88
classified information
  access to, 117–19
  government control of, 119–21
Clean Water Act, 121
Clemente, Michael, 119–20
Cleveland Plain Dealer, 91
Clinton, Hillary, 121
the cloud
  Fourth Amendment and, 124–25
  individual rights and, 121–23
  legal protections for, 123–24
  Privacy Protection Act and, 127–28
Code: And Other Laws of Cyberspace (Lessig), 122
Code of Judicial Conduct, 88
Colorado Supreme Court, on email privacy, 107
Columbus Ledger-Enquirer news, 90
Committee on Open Government, 138
Committee to Protect Journalists, 120
computerized information
  alternative information sources, 46
  creating life-long dossiers, 41
  Justice Brennan on, 8
  Justice Stevens on, 7–8
  privacy-related court cases and, 38–39
  private interests served by, 44–45
  promise of confidentiality and, 46–47
  public interests served by, 42–44
  Supreme Court cases regarding, 5–9
  threats to children in, 40–41, 44–45
  threats to privacy in, 9–10
computers
  advantages of, 3
  privacy interests and, 4
Conference of Chief Justice/Conference of State Court Administrators (CCJ/COSCA) guidelines
  for bulk and compiled record access, 72
  for electronic court record access, 69–70
confidentiality
  agency promises of, 46–47
  in mediation process, 137

conflict resolution. *See* dispute resolution
Connecticut Freedom of Information Commission, 139
constant connectivity, 146, 156–58
Constitution of the United States, 98
Consumer Privacy Bill of Rights, 54
content-based exemptions, to court record access, 72–73
contextual privacy, 75–77
Controlled Substance Act of 1972, 6–7
court records
　changes in access to, 67–68
　electronic access. *See* electronic court record access
courtroom proceedings
　access to, 85–87
　broadcasting, 87–91
Cowles Publishing Co. v. Kootenai County Board of County Commissioners, 107
Cross, Harold, 74
Cuillier, David, 161
cyberspace
　informational privacy and, 145–46
　laws for, 122

D

Dalglish, Lucy, 119–20
data controllers, 55
data protection
　description of, 51
　European approach to, 53–54
　in Spain, 51–52
　transparency balanced with, 58–59
　See also informational privacy
Davis, Charles N., 161–62
Davis, Paul v., 5
deep web (invisible web), 121
Denuo media consultancy, 147
Denver Publishing Co. v. Board of County Commissioners of County of Arapahoe, 107
Department of Air Force v. Rose, 12, 19–20
Department of Justice (DOJ), 119–20
Department of State v. Ray, 23–24

Department of State v. Washington Post, 21–22
Dershowitz, Alan, 128
Digital Due Process coalition, 99
digital eyeglasses, 150
Dimora, Jimmy, 91
disclosure decision model, 148–49
disclosure, serving public interests, 42–44
dispute resolution
　administrative adjudication in, 138–39
　mediation in, 136–37
　negotiated rulemaking in, 140–41
　negotiation in, 135–36
　ombuds offices for, 137–38
　online open portals for, 141–43
　in privacy vs. transparency issues, 133–35
Doctrine of FOIA Privacy Exceptionalism
　creation of, 29–31
　Supreme Court cases involving, 20–26
Dodgeball location-based service, 149
DOJ (Department of Justice), 119–20
Dooley, John, 70
Downie, Leonard, Jr., 120
Duhigg, Charles, 121

E

Easton Area School District v. Baxter, 106
ECJ (European Court of Justice), 52
ECPA Amendments Act, 126
EFOIA (Electronic Freedom of Information Act), 18
E-Government Act of 2002, 69
Elastic Compute Cloud (EC2), 121
electronic communication service (ECS), 125–26
Electronic Communications Privacy Act (ECPA)
　Digital Due Process coalition and, 99
　protection of/amendments to, 126
electronic court record access
　of bulk and compiled records, 71–72
　categorical exemptions to, 72–74
　CCJ/COSCA guidelines for, 69–70
　development of, 68
　neutrality and, 74–75

PACER system and, 69
remote access/jammie surfers and, 70–71
Electronic Freedom of Information Act
  (EFOIA)
  amendments to, 67
  extending neutrality, 75
  requirements of, 18, 104
electronic media
  in federal courts, 87–91
  in state courts, 91–94
electronic surveillance
  journalism and, 121–23
  social media and, 156
Ellis, Daniel, 150
email privacy
  federal/state FOI statutes and, 104–6
  government transparency vs., 97
  judicial interpretations regarding, 106–9
  in workplace, 97, 101–4
Erdos, David, 61
Estes v. Texas, 88
EU (European Union), 52–53
Europe
  communication technologies
    and, 153–54
  conception of privacy in, 53–54
European Convention on Human
  Rights, 53, 153
European Court of Human Rights, 54
European Court of Justice (ECJ), 52
  on freedom of expression and privacy,
    56–58
  on transparency and privacy, 58–59
European Federation of Journalists, 56
European Newspaper Publishers' Association, 56
European Union (EU), 52–53
European Union (EU) Charter of
  Fundamental Rights, 53–54
European Union (EU) Data Protection
  Directive
  journalistic exemptions in, 56–58
  overview of, 54–56
  transparency and data privacy in, 56–58
European Union (EU) Data Protection
  Regulation
  controversy regarding, 52–53

journalistic exemptions in, 56–58
overview of, 55–56
"right to be forgotten" in, 59–61
transparency and data privacy in, 56–58

F

Facebook
  investments in, 154–55
  personal information and, 145, 147
  privacy policies, 153
Falmouth Fire Fighters' Union Local 1497 v.
  Town of Falmouth, 103–4
Family Education Rights, 133
Favish, National Archives and Records
  Administration v., 24–26
Federal Bureau of Investigation (FBI), 9
Federal Bureau of Investigation v. Abramson, 20
federal courts, broadcasting from, 87–91
federal government online portal, 142
Federal Rules of Criminal Procedure,
  84, 88, 89
Ferreira, Russell, 104
First Amendment
  court access and, 85–87
  privacy rights vs., 54, 133–35
The First Amendment is an Information Policy
  (Balkin), 122
Fitzgerald, John W., 36, 37
Fleischer, Peter, 53
Flood, Daniel J., 22
Florida
  Constitution, 133
  public records mediation program, 136–37
Florida, Chandler v., 88–89
Florida v. City of Clearwater, 107
FOIA (Freedom of Information Act). *See*
  Freedom of Information Act (FOIA)
Foster, Vincent, Jr., 24–26
Fourth Amendment
  cloud protection and, 124–25
  privacy issues and, 98–99
  protection of emails and, 99–101
Fox News, 119–20

freedom of expression
    cyberspace and, 122
    government secrecy and, 117–18
    informational privacy and, 56–58
    in journalism, 121–23
Freedom of Information Act (FOIA)
    balancing test of, 19–20
    burden of proof for, 24–26
    central purpose standard for, 22–23
    disclosure mandate of, 17
    email privacy and, 104–9
    exemptions within, 18–19
    federal agency compliance (study), 27–29
    multi-agency online portal for, 142
    neutrality and, 74–75
    noncompliance with, 17–18
    presidential application of, 26–27
    presumption of legitimacy rationale and, 23–24
    privacy exceptionalism and, 20–21
    public-record information and, 9, 11–12
    Supreme Court's control over, 29–31
Freedom of Information Advisory Council, 138
Freedom of Information Commission (Connecticut), 139
frictionless sharing, 156
Frons, Marc, 118

G

Galison, Peter, 119
Gardephe, Paul G., 103
Gates, Paul, 165–66
gender-reveal videos, 150–51
Georgia First Amendment Foundation, 135–36
Global Positioning System (GPS) capabilities, 155–56
Globe Newspaper Co. v. Superior Court, 86
Goldsmith, Jack, 153
Goodale, James, 119–20
Google, 121, 123, 125
government transparency
    computerized databases and, 37–42
    email privacy vs., 97

FOIA and, 17
personal privacy and, 19–20
President Obama and, 16, 119
secrecy vs., 117–18
GPS (Global Positioning System) capabilities, 155–56
Graf, Daniel, 149
Greenberg, Andy, 118, 123
The Guardian newspaper, 117–18

H

Halstuk, Martin E., 162
Harlan, John Marshall, 88, 98
Hastings, United States v., 89
haul videos, 149
Hauptmann, Bruno, 88
Healthcare Information Portability and Accountability Act, 133
Holder, Eric, 26–27
Hollingsworth v. Perry, 90
Holman v. Superior Court of San Diego County, 108–9
Holmes v. Petrovich Development Co., 103
Horton, Frank, 4
hyperdissemination, 70

I

Idaho Supreme Court, on email privacy, 107
Illinois
    Public Access Counselor in, 138
    ruling on personal information redaction, 45–46
imaginary audience, in adolescence, 149
Indiana, Public Access Counselor in, 138
informational privacy
    cyberspace and, 122, 145–46
    freedom of expression and, 56–58
    as fundamental human right, 53–54
    international concerns regarding, 51–53
    necessary disclosure vs., 18–19
    right to control of, 10

Supreme Court cases regarding, 5–9
transparency and, 58–59
See also data protection
In re Reserve Fund Securities and Derivative Litigation v. Reserve Management Co., 103
In re Silberstein, 109
Instagram, 154
invisible web (deep web), 121
Iowa, ombuds office in, 138

J

jammie surfer problem, 70–71
Jones, United States v., 99
journalistic exemptions, in Data Protection Regulation/Directive, 56–58
journalists
    emerging technologies and, 121–22
    legal protections for, 123–28
    security skills and, 123
    using government sources, 119–21, 122–23
Judicial Conference of the United States, 89
judicial proceedings
    for access-specific dispute resolution, 139
    access to, 85–87
juvenile proceedings, right of access in, 87

K

Kansas courts
    on computer privacy issues, 44, 45, 105
    on electronic media in courts, 92
Katz v. United States, 6, 98
Kennedy, Anthony, 25–26
Kim, Stephen J., 119
Kiriakou, John, 119
Koops, Bert-Jaap, 59–60
Kootenai County Board of County Commissioners, Cowles Publishing Co. v., 107
Kozinets, Peter S., 106–9

L

Lake County Sheriff's Department, Ohio ex rel. Wilson-Simmons v., 106
Land, Clay, 84
law enforcement privacy Exemption 7(C)
    in denial of FOIA requests, 27–29
    Supreme Court exploitation of, 30–31
Leahy, Patrick, 126
legal protections
    Electronic Freedom of Information Act as, 126
    emerging technology and, 123–24
    Fourth Amendment as, 124–25
    Privacy Protection Act as, 127–28
    Stored Communications Act as, 125–26
legislative exemptions, to court record access, 73
Leibowitz, Shamai, 119
Le Monde, 154
Lessig, Lawrence, 122
lifelogging, 150
Lindqvist, Bodil, 56–58
Lioi, Sara, 91
LiveJournal, 147
local protocols, negotiation regarding, 140–41
Loving Care Agency, Stengart v., 102–3

M

MacKinnon, Rebecca, 122
Manning, Chelsea, 119
Mann, Steve, 150
Marten, J. Thomas, 84, 90
Massachusetts Supreme Court, on computer privacy issues, 40
Mayer-Schönberger, Viktor, 59–60
McNealy, Scott, 145
Mechling, Meredith, 109
Mechling v. City of Monroe, 109
mediation, for conflict resolution, 136–37
Medico, Charles, 22
Memorandum on the Freedom of Information Act, 26

Memorandum on Transparency and Open
    Government, 26
Michigan Supreme Court
    on computerized data issues, 38–39
    on disclosure serving public interests, 43
Miller, Benjamin, 45–46
Miller, United States v., 5–6
Mill, John Stuart, 152–53
Minnesota courts, electronic media in, 94
Minnesota Supreme Court
    on computer privacy issues, 45
    on court record access, 70–71
Missouri courts, electronic media in, 93
Moreno, Federico, 91
Morozov, Evgeny, 122
Muse, Christopher J., 104

N

Narcissism Personality Inventory, 148
narcissism, social media and, 148
*National Archives and Records Administration v.
    Favish*, 24–26
national data center, controversy
    regarding, 4
National Security Agency (NSA), 117, 120–21
negotiated rulemaking, 140–41
negotiation, in conflict resolution, 135–36
neutrality
    contextual privacy and, 77
    in FOIA revolution, 74–75
New Hampshire Supreme Court,
    on disclosure, 43
New Jersey Supreme Court
    on computerized data issues, 39–40
    on court record access, 71
    on disclosure serving private interests, 44
    on workplace email privacy, 102–3
New Mexico
    court rulings on disclosure, 43
    online open portals in, 142
New York
    Committee on Open Government, 138
    courts, electronic media in, 94

New York magazine, 146
New York Times, 29, 118–21, 123
Niedzviecki, Hal, 148
Nisbet, Miriam, 138
Nissenbaum, Helen, 76
North Carolina courts, electronic
    media in, 93
Norton, David C., 91
NSA (National Security Agency), 117, 120–21

O

Obama, Barack
    applying FOIA, 26–27
    FOIA denials and, 27–30
    government transparency and, 16,
        119, 129
    prosecution of leakers by, 119–21
O'Connor, Sandra Day, 20
ODR (online dispute resolution) system, 142
Office of Government Information Services
    (OGIS), 104, 134, 136, 138
Office of Open Records, 134, 137
*Ohio ex rel. Wilson-Simmons v. Lake County
    Sheriff's Department*, 106
Ohio Supreme Court
    on computerized information about
        children, 40–41
    on disclosure serving private interests,
        44–45
    on FOI and email privacy, 106
Oklahoma Supreme Court
    on computerized information, 40
    on court record access, 70
    on disclosure serving public interests, 43
Olmstead v. United States, 152
ombuds offices, for dispute resolution, 137–38
On Liberty (Mill), 152–53
online dispute resolution (ODR) system, 142
online open portals, for public record access,
    141–43
Open Government Ombudsman, 136
Open Meetings Law Enforcement Team
    (OMLET), 136

Openness Promotes Effectiveness in our National (OPEN) Government Act, 104
operational security, journalism and, 123
Ostapowicz, Kitty, 146–47

P

PACER (Public Access to Court Electronic Records) system, 69, 71
paper records, computerized information vs., 40–42
Patino, Michael, 100–101
Patino, Rhode Island v., 100–101
Paul v. Davis, 5
Peltz-Steele, Richard J., 163–64
Pennsylvania courts, on email privacy, 106, 109
Pennsylvania, open records mediation in, 137
personal data
    definition of, 55
    redaction of, 45–46
personal privacy
    controversy regarding, 4
    FOIA protections for, 18–19
    government transparency and, 19–20
    "practical obscurity" doctrine and, 9–10
    public interests vs., 42–44
    redaction protecting, 45–46
    right to control of, 10
    technology and, 145–46
    See also privacy
personal privacy exemption, 6
    in denial of FOIA requests, 27–29
    Supreme Court exploitation of, 30–31
Peters, Jonathan, 165
Petrovich Development Co., Holmes v., 103
Pew Internet & American Life Project, 85
Plinky, 149
PPA (Privacy Protection Act), 127–28
"practical obscurity" doctrine, 4, 9
Press-Enterprise Co. v. Riverside County Superior Court, 86
press freedom, control of information vs., 119–21
presumption of legitimacy rationale, 23–24

privacy
    computers and, 38–39 *See also* computerized information
    connectivity and, 156–58
    contextual, 75–77
    definitions of, 37–38, 39
    European attitudes toward, 53–54, 153–54
    Fourth Amendment and, 98–99
    individual rights and, 10
    public records and, 11–12
    as right to be let alone, 151–52
    technological advances vs., 145–46
    transparency vs., 133–35
    in workplace emails, 97, 101–4
    See also personal privacy
Privacy Act of 1974, 4, 11, 55, 76, 133
Privacy and Freedom (Westin), 10, 18
privacy exceptionalism thesis
    establishment of, 21–22
    expansion of, 23–24
    fortification of, 22–23, 30–31
Privacy Protection Act (PPA), 127–28
private interests, disclosure and, 44–45
Proposition 8, 90
public access laws
    conflict regarding, 134–35
    dispute resolution for, 135–40
    online open portals and, 141–43
Public Access Ombudsman, 134, 138
Public Access to Court Electronic Records (PACER) system, 69, 71
Public Information Board, 138
public interests, disclosure and, 42–44
public-record information
    defined, 11
    emails as, 105
    privacy interests vs., 11–12, 30–31
Pulaski County v. Arkansas Democrat-Gazette, 107

Q

quasi-ombuds model, 137–38
Quon, City of Ontario v., 99–100

R

Radio Television Digital News Association, 92
Ray, Department of State v., 23–24
RCS (remote computing service), 125–26
redaction of personal information, 45–46
Reding, Vivane, 60
Rehnquist, William
    anti-disclosure ruling by, 21, 22
    on computer technology, 3
    Paul v. Davis case, 5
relational-privacy rights, 25
remote computing service (RCS), 125–26
remote online access, to court records, 70–71
Reporters Committee for Freedom of the Press, U.S. Department of Justice v.
    disclosure of public records and, 11–12
    personal privacy and, 10
    "practical obscurity" doctrine and, 4, 9–10
    public vs. private information and, 11
Reserve Management Co., In re Reserve Fund Securities and Derivative Litigation v., 103
Rhode Island v. Patino, 100–101
Richmond Newspapers v. Virginia, 85–86
right of access
    to court records, 67–70
    to judicial proceedings, 85–87
"right to be forgotten"
    controversy regarding, 53
    implementation of, 60–61
    overview of, 51
    three concepts of, 59–60
    See also data protection; informational privacy
"right to oblivion", 59–60
The Right to Privacy (Warren & Brandeis), 152
Risen, James, 119
Riverside County Superior Court, Press-Enterprise Co. v., 86
Roe, Whalen v., 6–7
Rose, Department of Air Force v., 12, 19–20
Rosen, James, 119–20
Rosen, Jeffrey, 122
rulemaking, negotiated, 140–41
Rusbridger, Alan, 117
Ryan, James, 39–40, 43

S

Safe Drinking Water Act, 121
Safe Harbor Agreement, 55
Sartre, Jean-Paul, 156
Satamedia case, 57–58
Saturday Evening Gazette, 152
Savage, Judith, 100–101
SCA (Stored Communications Act). *See* Stored Communications Act (SCA)
Schakne, Robert, 9, 22
Schill v. Wisconsin Rapids School District, 108
secrecy paradigm, 76, 99
Securities and Exchange Commission (SEC), 103
self-disclosure, social media and, 148
Senat, Joey, 162–63
sexting by minors, 151
Shelnutt, United States v., 90
"show-cause" hearings, 86
Silberstein, Kenneth M., 109
Silver, Derigan, 164
Snowden, Edward, 117–18, 119, 120
social media
    addictiveness of, 148
    as business, 154–55
    consequences of, 156–58
    popularity of, 147–48
    surveillance capabilities of, 155–56
    variety of, 149–50
Solove, Daniel J., 76, 99
Sophos researchers, 147
Sotomayor, Sonia, 76, 99
Spanish Data Protection Authority, 51
spirit of democracy
    electronic surveillance and, 121–23
    regulating technology and, 123–24
Splichal, Sigman L., 162
Standing Committee on Rules of Practices and Procedures, 92
Stanford Law Review Online, 52
Starr, Kenneth, 10
state courts
    on alternative information sources, 46
    broadcasting from, 91–94

on computerized data issues, 38–42
on disclosure serving private interests, 44–45
on disclosure serving public interests, 42–44
on redaction of personal information, 45–46
State Records Committee (Utah), 139
Stengart v. Loving Care Agency, 102–3
Sterling, Jeffrey, 119
Steve Jackson Games, Inc. v. United States Secret Service, 127
Stevens, John Paul
on computer privacy issues, 7–8
on concept of privacy, 38
on control of personal information, 10
on disclosure of public records, 11–12
on FOIA's central purpose, 22, 43
"practical obscurity" doctrine and, 9–10
on presumption of legitimacy, 23–24
on public vs. private information, 11
Stevenson, John Paul, 85
Stewart, Daxton R. "Chip", 165
Stickam, 147
Stored Communications Act (SCA)
limitations of, 97, 99
protection provided by, 125–26
Sunshine Portal, 142
Superior Court, Globe Newspaper Co. v., 86
Superior Court of San Diego County, Holman v., 108–9
Supreme Court. *See* U.S. Supreme Court
surveillance capabilities, of social media, 155–56
Swift, Jonathan, 145–46
Sylvester, Ron, 84, 90

T

Tancer, Bill, 147
Taylor, Richard, 85
technology, privacy and, 145–46
See also social media
teenagers, social media and, 146–51

television cameras
in federal courts, 87–91
in state courts, 91–94
Texas, administrative adjudication in, 139
Texas, Estes v., 88
Thierer, Adam, 52–53
third-party doctrine
cloud protection and, 124–25
of Fourth Amendment, 98–99, 101
This Machine Kills Secrets (Greenberg), 118
Thoreau, Henry David, 154
Tobaccowala, Rashid, 147
Town of Falmouth, Falmouth Fire Fighters' Union Local 1497 v., 103–4
Toxic Waters series, 121
transparency
data privacy balanced with, 58–59
privacy vs., 133–35
Trilling, Lionel, 156
Twenge, Jean M., 148
Twitter
in federal courts, 90–91
personal information and, 145
in state courts, 92–94

U

Uniform Information Practices Act, 134
United States, Katz v., 6
United States, Olmstead v., 152
United States Secret Service, Steve Jackson Games, Inc. v., 127
United States v. Hastings, 89
United States v. Jones, 99
United States v. Miller, 5–6
United States v. Shelnutt, 90
United States v. Warshak, 100
U.S. Bureau of the Budget, 4
U.S. Circuit Courts of Appeals
on electronic media in courts, 91
on protection of emails, 100
U.S. Department of Justice v. Reporters Committee for Freedom of the Press
disclosure of public records and, 11–12

personal privacy and, 10, 38
"practical obscurity" doctrine and, 4, 9–10
public v. private information and, 11, 41, 43
U.S. Supreme Court
  computer privacy cases, 5–9
  control over FOIA, 29–31
  description of privacy, 38
  privacy jurisprudence and, 98–99
  privacy-related FOIA cases, 20–26
  protection of emails and, 99–101
Utah State Records Committee, 139
Utah Supreme Court, on electronic media in courts, 92

V

Virginia, Richmond Newspapers v., 85–86
Virginia's Freedom of Information Advisory Council, 138

W

Walczak, Witold, 151
Walker, Vaughn R., 84
Wall Street Journal, 29, 123
Warren, Samuel, 10, 151–52, 155
Warshak, United States v., 100
Washington Court of Appeals, on email privacy, 109
Washington Post, 120, 121
Washington Post, Department of State v., 21–22
web-based open portals, 141–43

Westbrook, Robert, 42
Westin, Alan, 10, 18
West Virginia Supreme Court, on email privacy, 107–8
Whalen v. Roe, 6–7
whistleblowers
  overview of, 117–19
  prosecution of, 119–21
Whitman, James, 53, 54
Wichita Eagle newspaper, 84, 90
WikiLeaks, 119, 123
Wired, 120
Wisconsin Rapids School District, Schill v., 108
Wisconsin Supreme Court, on email privacy, 108
workplace email privacy, 97, 101–4
Wu, Tim, 122, 153

Y

Youm, Kyu Ho, 164
young adulthood, social media and, 147–51
YouTube
  gender-reveal videos and, 150–51
  haul videos on, 149

Z

Zuckerberg, Mark, 154–55
Zupancic, Bostjan, 154

SUSAN DRUCKER
Series Editor

Acknowledging the variety of ways in which the disciplines of communication and law converge, the aim of this series is to publish books at the nexus of these two areas with particular attention paid to communication in law in the changing media landscape.

Utilizing both qualitative and quantitative methodologies, volumes in this series provide analysis of issues at the interdisciplinary and international level such as free and responsible speech, media law, regulation and policy, press freedoms and governance of new media.

For additional information about this series or for the submission of manuscripts, please contact Dr. Drucker at susan.j.drucker@hofstra.edu.

To order other books in this series, please contact our Customer Service Department:

(800) 770-LANG (within the U.S.)
(212) 647-7706 (outside the U.S.)
(212) 647-7707 FAX

Or browse online by series at www.peterlang.com.

www.ingramcontent.com/pod-product-compliance
Ingram Content Group UK Ltd.
Pitfield, Milton Keynes, MK11 3LW, UK
UKHW021326180426
11947UKWH00017B/1472